ENERGY POLICIES, POLITICS AND PRICES

T0295418

ALTERNATIVE ENERGY SYSTEMS IN BUILDINGS

ENERGY POLICIES, POLITICS AND PRICES

Additional books and e-books in this series can be found on Nova's website under the Series tab.

ENERGY POLICIES, POLITICS AND PRICES

ALTERNATIVE ENERGY SYSTEMS IN BUILDINGS

ANTONIO COLMENAR-SANTOS
ENRIQUE ROSALES-ASENSIO
AND
DAVID BORGE-DIEZ
EDITORS

nova
science publishers
New York

Copyright © 2019 by Nova Science Publishers, Inc.

All rights reserved. No part of this book may be reproduced, stored in a retrieval system or transmitted in any form or by any means: electronic, electrostatic, magnetic, tape, mechanical photocopying, recording or otherwise without the written permission of the Publisher.

We have partnered with Copyright Clearance Center to make it easy for you to obtain permissions to reuse content from this publication. Simply navigate to this publication's page on Nova's website and locate the "Get Permission" button below the title description. This button is linked directly to the title's permission page on copyright.com. Alternatively, you can visit copyright.com and search by title, ISBN, or ISSN.

For further questions about using the service on copyright.com, please contact:
Copyright Clearance Center
Phone: +1-(978) 750-8400 Fax: +1-(978) 750-4470 E-mail: info@copyright.com.

NOTICE TO THE READER

The Publisher has taken reasonable care in the preparation of this book, but makes no expressed or implied warranty of any kind and assumes no responsibility for any errors or omissions. No liability is assumed for incidental or consequential damages in connection with or arising out of information contained in this book. The Publisher shall not be liable for any special, consequential, or exemplary damages resulting, in whole or in part, from the readers' use of, or reliance upon, this material. Any parts of this book based on government reports are so indicated and copyright is claimed for those parts to the extent applicable to compilations of such works.

Independent verification should be sought for any data, advice or recommendations contained in this book. In addition, no responsibility is assumed by the publisher for any injury and/or damage to persons or property arising from any methods, products, instructions, ideas or otherwise contained in this publication.

This publication is designed to provide accurate and authoritative information with regard to the subject matter covered herein. It is sold with the clear understanding that the Publisher is not engaged in rendering legal or any other professional services. If legal or any other expert assistance is required, the services of a competent person should be sought. FROM A DECLARATION OF PARTICIPANTS JOINTLY ADOPTED BY A COMMITTEE OF THE AMERICAN BAR ASSOCIATION AND A COMMITTEE OF PUBLISHERS.

Additional color graphics may be available in the e-book version of this book.

Library of Congress Cataloging-in-Publication Data

ISBN: 978-1-53614-203-7

Published by Nova Science Publishers, Inc. † New York

CONTENTS

Preface		**vii**
Chapter 1	Passive Techniques and Clean Conditioning Systems *Francisco Mur-Pérez and Clara M. Pérez-Molina*	**1**
Chapter 2	Energy Consumption Reduction in Large Buildings *Lya Noemí Terán de Lober, Ursula Eicker and Carlos-Ignacio Cuviella*	**43**
Chapter 3	Passive Climatization in Hot Climates *Clara Pérez-Molina and Francisco Mur-Pérez*	**77**
Chapter 4	Solar Thermal Systems for High Consumption Demand Buildings *Julio Vale-Vale, Carlos de Palacio and Pedro-Miguel Ortega-Cabezas*	**113**
About the Editors		**143**
Index		**147**

PREFACE

This book constitutes the refereed proceedings of the 2018 International Conference on Alternative Energy Systems in Buildings, which was held on 23rd May 2018. 2018 International Conference on Alternative Energy Systems in Buildings intends to provide an international forum for the discussion of the latest high-quality research results in all areas related to Alternative Energy Systems in Buildings. The editors believe that readers will find following proceedings interesting and useful for their own research work.

This book contains the Proceedings of the 2018 International Conference on Alternative Energy Systems in Buildings held online (https://enriquerosales.wixsite.com/virtualconferences), on 21st May, 2018. It covers significant recent developments in the field of Alternative Energy Systems in Buildings from an applicable perspective.

ADVISORY BOARD

Organizing Committee Chair:
Enrique Rosales Asensio, PhD,
Departamento de Física, Universidad de La Laguna, La Laguna, Spain
Email: erosalea@ull.edu.es

PROGRAM COMMITTEE CHAIRS:

Enrique González Cabrera, PhD
Departamento de Ingeniería Química y Tecnología Farmacéutica, Universidad de La Laguna La Laguna, Spain
Email: eglezc@ull.edu.es

Antonio Colmenar Santos, PhD
Departamento de Ingeniería Eléctrica, Electrónica, Control, Telemática y Química Aplicada a la Ingeniería, Universidad Nacional de Educación a Distancia, Madrid, Spain
Email: acolmenar@ieec.uned.es

David Borge Diez, PhD
Departamento de Ingeniería Eléctrica y de Sistemas y Automática, Escuela Técnica Superior de Ingenieros de Minas de León, León, Spain
Email: dbord@unileon.es

SCIENTIFIC COMMITTEE

Manuel Castro-Gil, PhD, Universidad Nacional de Educación a Distancia, Madrid, Spain

José María Pecharromán Lázaro, ENDESA, Palma de Mallorca, Spain

Elio San Cristobal Ruiz, PhD, Universidad Nacional de Educación a Distancia, Madrid, Spain

África López-Rey, PhD, Universidad Nacional de Educación a Distancia, Madrid, Spain

Rosario Gil Ortego, PhD, Universidad Nacional de Educación a Distancia, Madrid, Spain

Salvador Ruiz Romero, ENDESA, Barcelona, Spain

Jorge Blanes Peiró, PhD, Universidad de León, León, Spain

May 2018

Editors

In: Alternative Energy Systems … ISBN: 978-1-53614-203-7
Editors: A. Colmenar-Santos et al. © 2019 Nova Science Publishers, Inc.

Chapter 1

PASSIVE TECHNIQUES AND CLEAN CONDITIONING SYSTEMS

Francisco Mur-Pérez* and Clara M. Pérez-Molina
Departamento de Ingeniería Eléctrica, Electrónica, Control,
Telemática y Química Aplicada a la Ingeniería,
Universidad Nacional de Educación a Distancia (UNED),
Madrid, Spain

ABSTRACT

Over the past few years, the use of passive cooling and heating technologies has become more common for reducing the energy consumption of buildings. However, these technologies are not often used for development cooperation building systems. In 2010, over 42.3 million people were forced to abandon their homes and live in temporary shelters. Buildings intended for children or the elderly are often climatized to improve the indoor thermal conditions. In this chapter, a cost reduction in climatization for this type of building based on passive systems is proposed and studied. Building site optimization is performed to improve

* Corresponding Author Email: fmur@ieec.uned.es.

thermal behavior. To achieve this, computational fluid dynamics tools have been used. The integration of these passive systems allows the peak power demand to be reduced by up to 50% and the yearly energy consumption to be reduced by approximately 40%. These reductions are studied for conventional and renewable energy systems, showing that passive systems provide better thermal comfort and reduce initial investment and energy consumption, making low-cost buildings feasible. Passive systems should be a fundamental focus of study in development cooperation buildings.

Keywords: low-cost housing, thermal comfort, international development

NOMENCLATURE

ASHRAE	American Society of Heating, Refrigerating and Air-Conditioning Engineers
CFD	Computational fluid dynamics
COP	Coefficient of performance
CR	Cool roof
DX	Direct expansion
EC	Fuel energy consumption in the generator (kWh)
EPS	Expanded polystyrene insulation
FCM	Fanger comfort model
FV	Finite volume
HVAC	Heating, ventilation, and air conditioning
IDC	Internally displaced camp
IDPs	Internally displaced persons
PDEs	Partial differential equations
PMV	Predicted mean vote
PPD	Predicted percentage dissatisfied
RANS	Reynolds-averaged Navier-Stokes
RH	Relative humidity
ROI	Return of investment

SEER	Seasonal energy efficiency rate
SR	Solar reflectance
UNHCR	Office of the United Nations High Commissioner for Refugees
ρ	Air density (kg/m^3)
η_{Aux}	Energy efficiency of the auxiliary devices (%)
η_{Gen}	Energy efficiency of the diesel generator (%)
C	Flow coefficient related to the size of the opening (m^3/sPa)
C_p	Wind pressure coefficient at the position of the surface
DP	Pressure difference across the opening (Pa)
e	Infrared emittance
FA	Zone floor area (m^2)
f_{cl}	Ratio of clothed surface to nude surface
h_c	Convective heat transfer coefficient (W/m^2K)
M	Metabolic rate (W/m^2)
N	Flow exponent
NP	Number of people
OC	Occupancy density ($people/m^2$)
P_w	Partial water pressure (Pa)
P_W	Surface pressure due to wind (Pa)
q	Volumetric flow through the opening (m^3/s)
$\dot{Q}_{min}$	Minimum fresh air per person (m^3/s)
$\dot{Q}_v$	Natural ventilation rate (m^3/s)
R	Thermal resistance (m^2K/W)
T_a	Air temperature (ºC)
T_{cl}	Clothing temperature (ºC)
T_r	Mean radiant temperature (ºC)
U	Coefficient of transmission (W/m^2K)
v_z	Mean wind velocity at height z (m^2/s)
W	External work (W/m^2) of the body area

INTRODUCTION

Due to different situations, both natural and anthropogenic, many people around the world live under degraded conditions as internally displaced persons (IDPs). According to the Office of the United Nations Commissioner for Refugees (UNHCR), IDPs [1] are people displaced within their own country due to conflict, natural disasters, or violations of human rights. This population is currently the largest population of concern to the organization. In 2010, some 27.5 million people were displaced internally by conflict or violence. Additionally, over 42.3 million were forced to abandon their homes owing to natural disasters. Many of these examples can be seen in Haiti, Pakistan, Afghanistan, Somalia, and Yemen. As the number of internally displaced persons (IDPs) continues to increase, there is a strong need to improve their living conditions. In 2011, international organizations protected and assisted over 17 million IDPs. In contrast to people who have not crossed borders in their relocation, many people currently live as refugees in a country different from their habitual living place. Those people, especially economically motivated migrants, travel around the world to improve their future prospects as well as those of their families. They do not receive any protection from their home country and usually live in very poor conditions. Natural disasters, such as the Japan Earthquake, also necessitate temporary shelters for the inhabitants until the country or city is reconstructed.

One of the primary purposes of UNHCR aid is to safeguard the rights and well-being of refugees and IDPs; however, the final objective is to help find durable solutions that will allow people to rebuild their lives. Temporary shelters are one of the primary resources in camps. Buildings should be intended for fast development and construction and should be easily built because no advanced building resources exist in these situations. The shelters must be durable, low cost, and as comfortable as possible. Due to a lack of facilities or energy, shelters are usually not climatized at all, leading to very uncomfortable conditions.

Passive climatization techniques are usually intended for energy consumption reduction in buildings but are not often involved in development cooperation. Passive systems can be used for entirely passive building designs or as a way to reduce energy consumption for conventional heating, ventilation, and air conditioning (HVAC) systems. The authors have previous research experience in the design and simulation of passive cooling technologies [2-5]. The aims of this chapter are to extend previous research to climatized shelters used for IDPs and evaluate the results. In isolated areas, buildings needing HVAC, such as hospitals or children's shelters, are climatized using electricity provided by diesel generators or any other fuel. Renewable energy systems based on solar thermal energy and absorption systems can also provide an effective solution because no fuel is required. In this chapter, a real internally displaced camp (IDC) in Central America is analyzed and a low-cost building is studied. The building is fully made of concrete, which makes it easy to build while providing high durability. To evaluate the cooling needs to improve thermal comfort, the energy consumption is studied for the base case and for seven additional cases combining cool roof (CR) technology, insulation, and natural ventilation. The energy consumption and the required cooling power are evaluated for both electrical cooling (direct expansion) and solar-driven absorption systems, proving that a small investment in low-cost passive techniques improves thermal comfort and reduces the initial investment in HVAC systems and yearly energy consumption. Passive systems can be used to increase comfort in shelters and are a profitable way to improve their thermal behavior in the future. Upon return to their countries of origin, both IDPs and former refugees may encounter insecurity, destroyed homes and infrastructure, and a lack of employment opportunities and basic services. The use of low-cost passive climatization systems to make temporary shelters yields buildings that can be effectively used in the future. To provide a full analysis and improve both the thermal and energy behavior of the building, a whole-building analysis, including energy and computational fluid dynamics (CFD), is used. EnergyPlus is used to this end because it allows whole-building analysis, and Design Builder software is used because it allows complex HVAC models and combined CFD analysis.

SIMULATION METHODOLOGY

Shelter Design Optimization

After an IDC is dismantled, the IDPs who were living there must return to their original residences or new buildings must be built in the IDC location to provide shelter with the highest level of comfort possible under the conditions. Unfortunately, several million displaced persons will never return to their country of origin or will live in an IDC for several years, including in Somalia, Yemen, Palestine, and many other countries. The development cooperation organizations, both private and public, usually finance these projects. In addition to funding, building techniques must be transferred to local builders to achieve the expected building quality. Due to a lack of specialists in building energy efficiency, techniques to reduce thermal demand are usually not used. Passive technologies usually require advanced building techniques and materials, which are not available in the studied situation. In this case, facilitating access to the aforementioned building systems is the only practical solution. To achieve the best results, the location and the constructive techniques are optimized along the study presented in this chapter. The proposed method is used for a case study in Central America (Haiti), where an earthquake caused catastrophic damage in buildings and infrastructures January 12th, 2010. However, this method can also easily be applied to other locations, including the area affected by the Japan Earthquake, as mentioned earlier. Haiti's capital, Port au Prince, was severely damaged by the 2010 earthquake, which dramatically reduced the country's infrastructure while causing a high number of IDPs. Over 100,000 buildings were destroyed, and an additional 200,000 were damaged to the point of needing to be repaired or rebuilt. Currently, a substantial housing deficit still exists. The earthquake caused approximately $4.3 billion in damage to the physical infrastructure. The destruction created approximately 2.3 million Haitian IDPs, who live in IDCs. Despite international cooperative programs, 1.3 million Haitians still live in tents or transitional shelters, producing a widespread landscape of scattered camps. These emergency settlements offer insecure living conditions and increase the risk of disease. The case studied here involves a new building location for small shelters.

Computational Fluid Dynamics (CFD) Analysis for Optimization

The building placement is very relevant to the energy thermal demand of the shelter and the indoor air quality. The building also must be optimized to improve thermal comfort if no HVAC systems are feasible and to reduce the energy consumption of both conventional and renewable energy systems. The analysis tools include comfort value analysis. These values are crucial because the main objective of the proposed methodology is find the most cost-effective way to achieve comfortable indoor thermal conditions.

External CFD analyses can be used to optimize building location. No cost is associated with optimizing the distribution, and strong benefits can be obtained. CFD is a family of numerical methods used to calculate the temperature, velocity, and other fluid properties throughout a spatial area, and it can also be used for the study of buildings. CFD allows air velocities, pressures, and temperatures inside the building and in the surroundings to be predicted. The obtained values can be used to calculate the thermal comfort inside the building. The method of finite volume (FV) can be effectively used for solving the set of partial differential equations involved in the model. PDEs describe the transport of momentum, energy, and turbulence quantities, as described in Eq. (1).

$$\frac{\partial(\rho\phi)}{\partial t} + div(\rho u\phi) = div(\Gamma\ grad\phi) + S \tag{1}$$

PDEs are converted into a set of simultaneous algebraic equations, and the space under analysis, both inside and outside of the building, is divided into a set of rectilinear cells, called a finite volume grid. The obtained algebraic equations are set up for each grid cell, and the whole set of equations is solved using numerical methods. The PDEs are discretized across the control volumes using linear profiles that represent the variation of the dependent variables between the grid points and their neighbors. The discretization for a one-dimensional finite volume grid is shown in Eq. (2), and the rewritten equation using a linear profile is shown in Eq. (3), where the subscript *P* represents the studied point and *W* and *E* represent the western and eastern adjacent points, respectively. The mean distance between *P* and *W* is represented by *w* (corresponding to a distance of $\frac{\delta(x)w}{2}$), and the mean distance between *P* and *E* is represented by *e* (corresponding to a distance of $\frac{\delta(x)e}{2}$).

$$(\rho u\phi)_e - (\rho u\phi)_w = \left(\Gamma\frac{\partial\phi_i}{\partial x_j}\right)_e - \left(\Gamma\frac{\partial\phi_i}{\partial x_j}\right)_w \tag{2}$$

$$\frac{1}{2}(\rho u)_e(\phi_E + \phi_P) - \frac{1}{2}(\rho u)_w(\phi_P + \phi_W) = \frac{\Gamma_e(\phi_E - \phi_P)}{(\partial x)_e} - \frac{\Gamma_w(\phi_P - \phi_W)}{(\partial x)_w} \quad (3)$$

The discretized equations are rearranged and take the form of a set of algebraic solutions that can be solved using numerical methods. These solutions are the fundamentals of the FV method utilized in the Design Builder (DB) software [6] external CFD analysis used in this chapter.

Both external and internal CFD analyses were performed.

External CFD Analysis Model

External CFD analysis was used to optimize the building site in the area. A TMY2-format file registered in the international airport near the location was used to simulate the weather conditions. This weather file provides contour conditions for calculations. For the analysis, a 1-k-e turbulence model was selected. This model is one of the most widely used and tested of all turbulence models and can be used reliably for most building air flows, including high-velocity flows and highly buoyant flows. This model belongs to the Reynolds-averaged Navier-Stokes (RANS) family of models, in which the instantaneous velocities in the Navier-Stokes equations and the energy equations are replaced with a mean and a fluctuating component. The resulting equations give rise to additional terms known as Reynolds stresses and turbulent heat flux components. Reynolds stresses are replaced by terms involving instantaneous velocities, where molecular viscosities are substituted for effective viscosities, and an analogous substitution is made in the energy equation. The effective viscosity is the sum of the molecular viscosity and a turbulent viscosity, which is derived from the turbulence kinetic energy and the dissipation rate of the turbulence kinetic energy. This approximate model is criticized for overestimating diffusion in the case of highly separating flows and has limitations for the prediction of jet spread. However, the standard k-e model is considered the most robust and reliable of all turbulence models, and the application of CFD to building analyses is accurate enough because of the absence of complex phenomena, such as hypersonic flows, reacting flows, superfluid flows, or shock waves. A non-uniform grid was used to add grid exceptions in the building corners, which ensures that the corners are located at the intersections of the grid to guarantee the accuracy of the results.

Internal CFD Analysis

Internal CFD analysis was used to evaluate the comfort parameters inside the building. Thermal comfort is evaluated in DB using thermal comfort evaluation parameters. The Fanger comfort model (FCM) is also evaluated. The FCM was published first in 1967 and then again in 1972 [7-9] and helped set the stage for the other two models, the Pierce two-node model and Kansas State University two-node model [10-12]. The mathematical model developed by Fanger is the best known of the three models and the easiest to use, as it has been presented in both chart and graph form [13, 14]. The environmental variables that influence the conditions of thermal comfort include environmental and physiological variables [15-17] (see Table 1).

Table 1. Variables used in comfort analysis

Environmental variables	Physiological variables
Air Temperature (T_a)	Skin Temperature (T_{sk})
Mean Radiant Temperature (T_r)	Core or Internal Temperature (T_{cr})
Relative air velocity (v)	Sweat Rate
Water vapor pressure in ambient air (P_a)	Skin Wettedness (w)
	Thermal Conductance (K) between the core and skin.

Using these parameters, the predicted mean vote (PMV) and predicted percentage dissatisfied (PPD) indexes are stated [18, 19] The CFD results are used to calculate the indexes and for comfort evaluation. According to the ISO 7730 norm, the PPD should not exceed 10%. PMV and PPD provide an accurate measure of people's response to the thermal comfort. Fanger realized that the predicted vote was only the mean value to be expected from a group of people and extended the PMV to predict the proportion of any population who will be dissatisfied with the environment. A person's dissatisfaction was defined in terms of their comfort vote. Those who voted outside the central three scaling points on the American Society of Heating, Refrigerating and Air-Conditioning Engineers (ASHRAE) scale were classified as dissatisfied [20-22]. The PPD is defined in terms of the PMV and adds no information to that already available in PMV [23-25]. The distribution of PPD is based on observations from climate chamber experiments, not field measurements [26]. However, the combination of both parameters is optimal for comfort evaluation because the PPD indicates whether the environment is acceptable and the PMV indicates whether the temperature is too hot or too cold [27, 28]. The PMV is obtained from the CFD results using Eq. 4, and the results are used to estimate the PPD using Eq. 5 [29-32].

$$
\begin{aligned}
PMV = (0.303e^{-0.036M} + 0.028) \\
\cdot \{(M - W) - 3.05E^{-3}[5733 - 6.99(M - W) - P_w] - 0.42[(M - W) - 5815] \\
- 1.7E^{-5}(5867 - P_a) - 0.0014M(34 - T_a) - 3.96E^{-8} \\
\cdot f_{cl}[(T_{cl} + 273)^4 - (T_r + 273)^4] - f_{cl}h_c(T_{cl} - T_a)\}
\end{aligned}
\tag{4}
$$

- M is the metabolic rate (W/m^2)
- W the external work (W/m^2) of the body area
- P_w is the partial water pressure (Pa)
- T_a is the air temperature (ºC)
- f_{cl} the ratio of clothed surface to nude surface
- T_{cl} the clothing temperature (ºC)

- T_r the mean radiant temperature (ºC)
- h_c the convective heat transfer coefficient (W/m^{2}K)

$$PPD = 100 - 95e^{-(0.03353PMV^4 + 0.2179PMV^2)} \tag{5}$$

Indoor Air Quality and Ventilation Requirements

Ventilation is used to provide the building with external fresh air as well as adequate air quality. The proposed model uses mixed-mode cooling, defined as a scenario in which occupants can open windows and that is designed with effective passive strategies for limiting the effects of the external climate. The passively designed building is utilized to provide acceptable conditions for most of the year and is supplemented by a mechanical system on either an 'as and when required' basis or a seasonal basis [33]. The type of air flow was calculated for the shelter site using the minimum fresh air requirements per person [34]. The air flow rate is established using Eq. 6 and Eq. 7:

$$\dot{Q}_v = \frac{\dot{Q}_{min} \cdot NP}{1000} \tag{6}$$

$$NP = OC \cdot FA \tag{7}$$

where $\dot{Q}_v$ is the natural ventilation rate in (m^3/s), $\dot{Q}_{min}$ is the minimum fresh air per person (m^3/s), NP is the number of people, OC is the occupancy density (people/m^2), and FA is the zone floor area (m^2). The windows in the building are used for natural ventilation, and an additional opening is added in the roof to increase air flux. The ventilation rate (q) through each opening is calculated based on the pressure difference using Eq. 8, where q is the volumetric flow through the opening (m^3/s), DP is the pressure difference across the opening (Pa), n is the flow exponent ranging from 0.5 for fully turbulent flow to 1.0 for fully laminar flow, and C is the flow coefficient related to the size of the opening (m^3/sPa). The wind-driven ventilation is calculated for the building openings, and TMY2 meteorological data are used. The pressure at any point on the surface of a building facade can be represented by Eq. 9.

$$q = C \cdot (DP)^n \tag{8}$$

$$P_W = 0.5 \cdot \rho \cdot C_p \cdot v_z^2 \tag{9}$$

P_W is the surface pressure due to wind (Pa), ρ is the air density (kg/m^3), C_p is the wind pressure coefficient at the surface, and v_z is the mean wind velocity at height z (m^2/s).

HVAC Systems

Due to the lack of energy facilities in IDCs, two different HVAC systems available for isolated use are studied: a system based on an electrical generator using fuel, which is the most common system in IDP camps, and a system based on renewable energy. In the case study located in Haiti, no energy systems exist in the IDC, fitting the proposed model.

The first HVAC system is a direct expansion (DX) system using electricity as driving energy. The DX system is used to maintain the required temperature. Operative temperature control is the most effective control strategy to obtain higher comfort values, but it is not a realistic situation in emergency shelters.

The energy consumption for the direct expansion is calculated using the seasonal energy efficiency rate (SEER) as stated in Eq. 10.

$$EC_{DX} = \frac{Q_{cooling}}{SEER \cdot \eta_{Gen} \cdot \eta_{Aux}} \tag{10}$$

where EC is the fuel energy consumption in the generator (kWh), η_{Gen} is the energy efficiency of the diesel generator, and η_{Aux} is the energy efficiency of the auxiliary devices.

For the renewable energy strategy, a solar-driven absorption machine is evaluated as a cooling system. The system allows cooling generation using thermal energy produced by solar collectors. The solar thermal energy required is calculated for the absorption system using Eq. 11, where the coefficient of performance (COP) is the absorption system COP and η_{Th} is the energy efficiency of the thermal storage and distribution system.

$$EC_{ST} = \frac{Q_{cooling}}{COP \cdot \eta_{th}} \tag{11}$$

Based on an hourly calculation, energy demand is studied for both cases, and a sensibility analysis is carried out. This study compares the energy consumption for the fuel-driven energy generator and the absorption system for different building strategies.

Building Internal Layout and Building Materials

Due to the need for a low-cost construction system, a high-occupancy building is studied. Usually in IDCs, 15 persons live in approximately 20 m^2. The thermal characteristic of the tents, constructed using textile materials, lead the temperature to increase beyond comfortable levels. In Haiti, where the study is carried out, the internal temperature in the tents can increase to 36ºC with 80% relative humidity (RH). These conditions are extremely uncomfortable and lead to disease proliferation and low quality of life for the IDP. The aim of the buildings is to provide a secure shelter for the IDP and to be the basis for possible future development of a new city. In the location selected for the study, as is characteristic of such cases, there are no advanced building techniques available. Therefore, an 'easy-to-build' building design and low-cost building material must be considered.

The Haitian government is managing the building of shelters to accommodate IDPs. Shelters are now being built without taking into account the possibility of using passive climatization techniques, which would reduce energy consumption in climatized facilities. Shelters intended for the elderly, children, and the ill always need HVAC systems. Temporary hospitals also require HVAC systems. The standard building rate is now 2.8 m^2 per person. According to this rate, a shelter for 10 persons is studied (Figure 1).

The occupancy levels will vary during the day. People will occupy the building to sleep during the night, but during the daytime, the mean occupation will be the half the nighttime occupancy. The occupation values are used to calculate the internal gains and the ventilation air requirements (Figure 2 and Table 2).

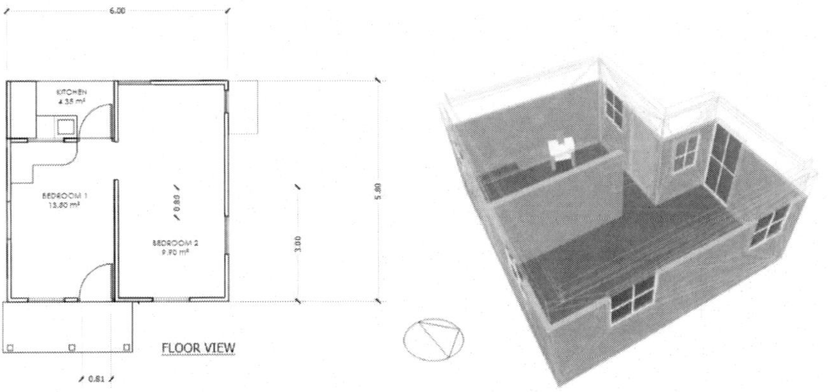

Figure 1. Building plant and indoor distribution used in simulation.

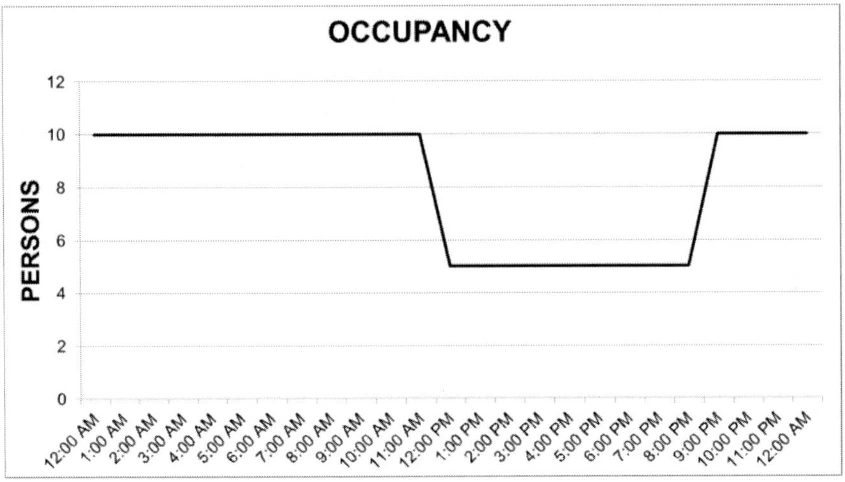

Figure 2. Building occupancy; Occupancy pattern proposed in the model.

Table 2. Occupancy during the day used in the simulation

Time	Density (people/m^2)
12PM - 9PM	0.18
9PM - 12 AM	0.36

The high occupancy levels require continuous air renewal to achieve the required indoor air quality. To this end, a mixed-mode model is used. For both the DX and absorption systems, natural ventilation is used when the outdoor temperature is two degrees lower than the indoor temperature. In this situation, full window opening is used. If the comfortable temperature is not achieved, the windows are programmed to close and an auxiliary system is initiated. The external air for ventilation is provided by opening windows or roof ventilation. The natural ventilation model described in section 2.2 allows a quite accurate calculation and ventilation optimization.

Building Materials

Due to the lack of monetary resources in the area, low-cost construction techniques are required [35, 36]. The absence of electricity and energy at the site necessitates the use of building materials that are easy to place and do not require skilled manpower. Concrete is used as the only construction material; it is economical, with high specific resistance, and can therefore be used for improved shelter reliability and durability. Concrete can be transported and produced on site by adding water, which makes it easy to use. In IDCs, low-quality materials, such as bricks, are usually used [37], which instills buildings with very poor thermal behavior, especially in terms of air infiltration. The characteristics of the concrete used are shown in Table 3.

The windows are made of 4 mm simple glazing with a U value of U_{glaz} = 6 (W/m^{2}K) and an aluminum frame of U_{frame} = 5.7 (W/m^{2}K). The used materials provide a low-cost solution and can be bought in the local market in nearby countries. The doors are made of aluminum, with U_{door} = 2 (W/m^{2}K). These materials were selected based on the money available for the building construction.

Table 3. Concrete physical characteristics proposed for the building

	Light Weight Concrete (LWC)	Reinforced Concrete (RC)
Conductivity [W/mK]	1.8	1.8
Specific Weight [kg/m^3]	1,600	2,400
Specific heat [kg/m^3]	0.278	0.278
Emissivity	0.9	0.93

Cool Roofs

Cool roofs (CRs) have been continuously gaining interest over the last several years. CRs are used in many studies as an effective way to mitigate the negative effects of urban heat islands [38-41]. The increase in temperature in urban areas relative to the rural surrounding areas has been documented in many cities and referenced [42, 43]. Moreover, due to climate change, the number and intensity of heat episodes is likely increasing. In urban areas, CR is presented as a solution for reducing energy consumption. In development cooperation, thermal discomfort in extreme cases can lead to heat-related illness and even death [44]. The increases in urban temperatures increase the demand and peak demand of energy for cooling, also increasing the initial investment for shelter cooling [45-48]. A reduction in cooling demand will produce higher efficiency levels and reduce the initial investment, which is a critical parameter. A CR system is a roofing system that is characterized by high solar reflectance (SR) and high infrared emittance (*e*) [49, 50]. The SR can be defined as the ability to reflect solar radiation, which further implies a reduction of heat transfer to the building, whereas higher *e* values imply a faster release of absorbed heat in the form of infrared radiation [51]. The high values of both of these two properties result in a roof with a lower temperature relative to a similar surface with lower SR and *e* values. The use of a CR would result decrease the heat penetrating into the shelter. A large number of experimental and modeling studies have demonstrated the benefits of CRs [52-55]. These studies concern reducing energy consumption [56, 59], the heat island effect, and energy cost; however, the system has not been studied as a way to both improve thermal comfort and decrease investment cost in cooperation development buildings [60-63].

The advantages of CR are presented below:

- Reduction of building heat gain: the temperature of a CR typically increases by only a few degrees Celsius above ambient temperature during the day.

- Average savings on summertime air conditioning expenditures in conditioned buildings of 10%–40% depending on the building characteristics and use as well as climatic conditions.
- Improvement of thermal comfort conditions in buildings without air-conditioned.
- Reduction of peak electricity demand, resulting in equipment downsizing, a lower probability of power failures on extremely hot days, and financial savings on fuel. In the case of a building climatized with renewable energy systems, low initial investment is needed, which makes the use of these technologies more profitable.
- Enhancement of the life expectancy of the roof system, reducing maintenance expenses due to less UV degradation and thermal fatigue.
- Mitigation of the heat island effect by 1–2°C, as less heat is transferred to the surrounding air.
- Reduction of air pollution and CO_2 emissions.

All of the cited literature studies mention these advantages [64], but there is no reference at all to the advantages derived from the use of this technology in cooperation development cases. CR technology is well known and applied in the US, where there are measurement standards related to CR and their use is a part of the energy code in many states. Some organizations, such as the USEPA Energy Star and the US Cool Roof Rating Council, as well as many green building codes incentivize and promote their use. Other countries, especially those in Europe, currently do not use this technology as a standard. The US and EU are the main actors in cooperation development [49, 53, 54, 55]. The EU supported a project called CoolRoofs and has developed and implemented an action plan for the promotion of CRs in European countries. This knowledge could be transferred to building technologies in cooperation development as a usual practice [65].

Results and Discussion

The objective of the passive technologies studied in this chapter is to reduce energy consumption in the shelter and initial investment cost. Using passive techniques reduces peak loads for heating or cooling, which in turn reduces equipment sizing and consequently initial cost. In the case study, only cooling is required due to the local climate (Haiti). To reduce cooling demand, a combination of low-cost and easy-to-build systems are evaluated. The proposed systems are expanded polystyrene insulation (EPS) in the roof, an opening in the roof for natural ventilation improvement due to the chimney effect, and CR. For the roof, a light-weight concrete, with a density of 1,600 kg/m^3, is used. In the base case, a thickness of 120 mm is the only used material, whereas in the EPS cases, an intermediate 84 mm EPS layer is included. In the walls, a 100 mm layer of standard concrete, with a density of 2,400 kg/m^3, is used for all simulations. The R-value is 3.508 for the roof in the base case and 0.311 if EPS is used. The wall value is 0.193. The technologies are combined to provide a cheap solution that lowers energy consumption in HVAC systems and reduces initial investment. All the building materials were simulated in the software, and the building model was established. To simulate the thermal behavior, HVAC systems are also introduced in the model, which allows a whole-building analysis and evaluates the influence of the building materials on the indoor comfort. The indoor comfort parameters can be calculated using comfort CFD calculations in Design Builder.

External Layout Optimization

The proper placement of buildings in the camp can improve thermal ventilation without any cost as well as reducing the heat island effect and increasing thermal comfort. Using the DB external CFD module, two different cases were simulated and studied: the first using the planned regular layout, shown in Figure 3, and the second using a non-symmetric distribution along the building lots, shown in Figure 4.

The results of the CFD simulation are shown in Figure 5 for the regular distribution case, whereas the CFD simulation results are provided for the non-symmetrical case in Figure 6a and Figure 6b.

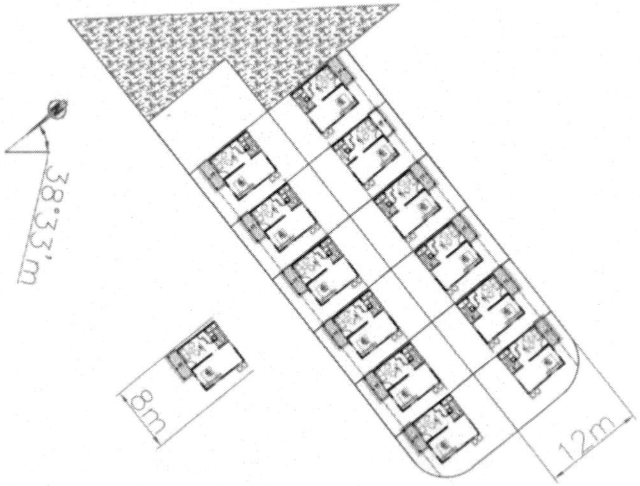

Figure 3. Regular plant distribution used for CFD simulation.

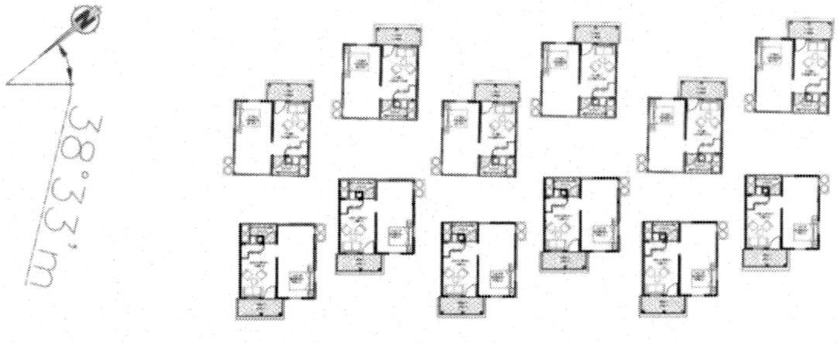

Figure 4. Non regular plant distribution used for CFD simulation.

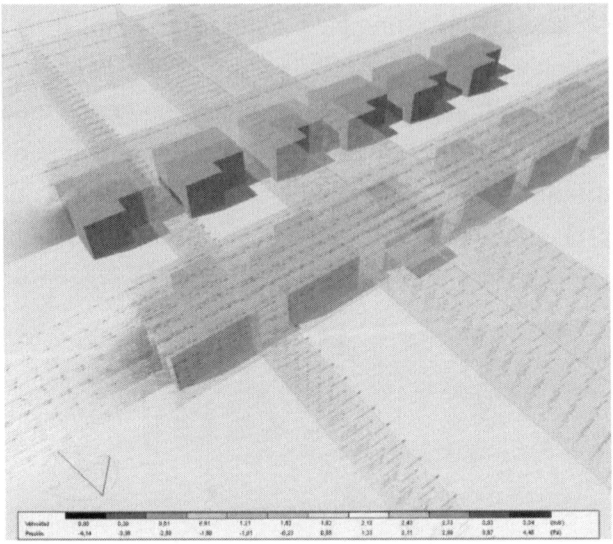

Figure 5. Regular plant distribution; external CFD analysis results.

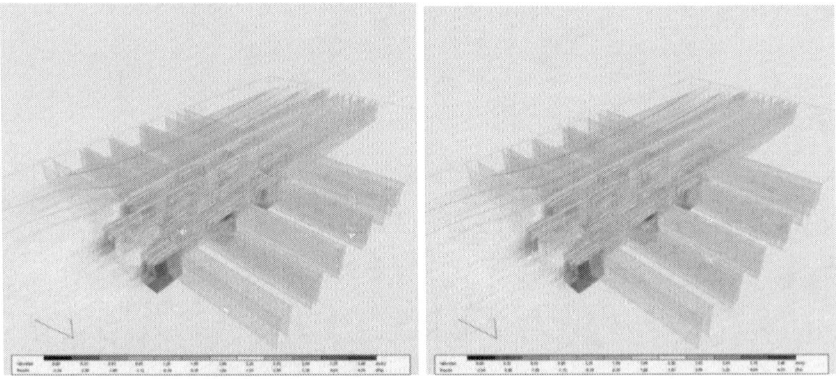

Figure 6a and 6b. Non regular plant distribution, external CFD analysis results.

The non-symmetric and non-regular distribution increases both the wind velocity and the pressure, as shown in the figures. The velocity values are increased by up to 25% for the intermediate shelters and over 57% for the buildings placed at the end of the lot. The velocity increase implies better performance in natural ventilation, reducing energy consumption in cooling systems. The velocity profile is optimal and pressure values facilitate natural ventilation in the mixed operation model.

According to Eq. 8 and Eq. 9, the cooling capacity by external air infiltration is higher when the pressure drop is greater and with increasing wind velocity.

Case Studies for Energy Demand Reduction

According to the presented strategy, eight building construction typologies were simulated for both regular and non-regular distributions. These cases are numbered as case 1 to case 8. For each case, two sub-cases with different HVAC systems are studied: the first using a DX HVAC system and second using a solar-driven absorption system. The main parameters for the DX and absorption cases are shown in Table 4 and Table 5, respectively. The proposed sixteen building models were simulated for both regular and non-regular distribution, in order to evaluate the distribution influence on energy demand and energy consumption. A total amount of 32 simulations were carried out. Due to the need for a low-cost and easy-to-build CR solution, a paint additive is used, which both reduces the conductivity of the original paint and increases its reflectivity. Acrylic water-based paint offers improved thermal behavior. Standard white acrylic paint is used, and the additive leads to a reflectivity of 0.9235 and a thermal conductivity of 0.1292 W/mK according to EN 12667. The cool roof materials are modeled using the software, and a mixed-mode model, including natural ventilation and an HVAC system, is used. When using the DX system, the cooling opening will not be opened. All of these detailed working conditions were established in the model, ensuring that the results correspond to the actual working mode and avoiding unrealistic situations, such as combined natural ventilation and roof opening.

Table 4. Case studies with DX HVAC system

Case	Ventilation and hvac	Roof
1DX	Windows and DX mixed mode	RC
2DX	Windows, roof opening and DX mixed mode	RC + Opening
3DX	Windows, and DX mixed mode	RC + CR
4DX	Windows, roof opening and DX mixed mode	RC + CR + Opening
5DX	Windows, and DX mixed mode	RC + EPS
6DX	Windows, roof opening and DX mixed mode	RC + EPS + Opening
7DX	Windows, and DX mixed mode	RC + EPS + CR
8DX	Windows, roof opening and DX mixed mode	RC + EPS + CR + Opening

Table 5. Case studies with solar driven absorption HVAC system

Case	Ventilation and hvac	Roof
1AB	Windows and solar driven absorption system mixed mode	RC
2AB	Windows, roof opening and solar driven absorption system mixed mode	RC + Opening
3AB	Windows, and solar driven absorption system mixed mode	RC + CR
4AB	Windows, roof opening and solar driven absorption system mixed mode	RC + CR + Opening
5AB	Windows, and solar driven absorption system mixed mode	RC + EPS
6AB	Windows, roof opening and solar driven absorption system mixed mode	RC + EPS + Opening
7AB	Windows, and solar driven absorption system mixed mode	RC + EPS + CR
8AB	Windows, roof opening and DX mixed mode	RC + EPS + CR + Opening

For the absorption system, fan-coils where chosen as the air distribution system. Other systems, such as a cooling ceiling, provide good solutions but also have high costs and risk of condensation due to the dew point temperature on cooled surfaces.

Cooling Demand for the Selected Day

The cooling power for both the DX and absorption systems must be sufficient to deliver the required cooling power for the hottest day of the year. For those conditions, the simulations were performed and the results were compared for the case studies. The best results are obtained for the combination of CR and improved ventilation using a roof opening, case 4 (Table 4 and Table 5) using non-regular distribution. Higher design day cooling demand was obtained for case 1 with regular distribution. Table 6 shows the results for case 1 (regular) and case 4 (regular and non-regular). Influence of external layout is shown for case 4 proving external layout influence on reducing cooling demand for the same building materials typology.

For regular distribution the cooling demand in the shelter decreases to 41% when using CR technology and adding a roof opening to improve natural ventilation. Comparing the results for case 1 (regular) and case 4 (non-regular), a 52% reduction is observed. For case 4 building typology a decrease of 11% is obtained at no cost changing plant layout, evidencing the need for studying and improving the shelter placement at the site. Figure 7 shows the comparison for design day results for case 1 (regular distribution) and case 4 (regular distribution) and Figure 8 shows the comparison for design day results for case 1 (regular distribution) and case 4 (non-regular distribution). Figure 9 shows the energy demand results for case 1 (regular distribution) and case 4 (non-regular distribution).

The cooling power decreases from 4.9 kW in the base case (case 1, regular distribution) to 2.40 kW in case 4 with a non-regular distribution. This finding implies a lower initial investment in the cooling system for the both DX and absorption systems. The cost reduction is extremely important in the absorption system because it also implies cost reduction in solar thermal collectors.

Table 6. Design day results for case 1 (regular) and case 4 (regular and non-regular distribution)

	Case 1 Regular												
	Windows	Walls	Basefloor	Interior walls	Roof	Doors and openings	Sensible cooling	Exterior infiltrations	Exterior Ventilation	Total cooling	Occupancy	Solar gains in windows	Sensible cooling
Specific demand	kW/m^2	kW/m^2	kW/m^2	kW/m^2	kW/m^2	kW/m^2	kW/m^2	kW/m^2	kW/m^2	kW/m^2	kW/m^2	kW/m^2	kW/m^2
	-0.0059	-0.0381	-0.0195	-0.0090	-0.0215	-0.0018	-0.1115	-0.0043	-0.0206	-0.1920	0.0267	0.0116	-0.1077
Net Cooling Power Demand	kW	kW	kW	kW	kW	kW	kW	kW	kW	kW	kW	kW	kW
	-0.1537	-0.9861	-0.5061	-0.2333	-0.5557	-0.0459	-2.8885	-0.1104	-0.5323	-4.9721	0.6916	0.3009	-2.7903
	Case 4 Regular												
Specific demand	kW/m^2	kW/m^2	kW/m^2	kW/m^2	kW/m^2	kW/m^2	kW/m^2	kW/m^2	kW/m^2	kW/m^2	kW/m^2	kW/m^2	kW/m^2
	-0.0045	-0.0086	-0.0288	-0.0111	-0.0050	-0.0028	-0.0522	-0.0038	-0.0183	-0.1131	0.0265	0.0116	-0.0499
Net Cooling Power Demand	kW	kW	kW	kW	kW	kW	kW	kW	kW	kW	kW	kW	kW
	-0.1165	-0.2223	-0.7465	-0.2869	-0.1292	-0.0718	-1.3530	-0.0982	-0.4732	-2.9291	0.6873	0.3009	-1.2934
Δ	0.0372	0.7637	0.2405	0.0536	0.4265	0.0259	1.5355	0.0123	0.0591	2.0429	0.0043	0.0000	1.4969
Δ % Case 1-Case 4 Regular	24%	77%	48%	23%	77%	57%	53%	11%	11%	41%	0%	0%	54%

	Case 4 Non Regular												
	Windows	Walls	Basefloor	Interior walls	Roof	Doors and openings	Sensible cooling	Exterior infiltrations	Exterior Ventilation	Total cooling	Occupancy	Solar gains in windows	Sensible cooling
Specific demand	kW/m^2	kW/m^2	kW/m^2	kW/m^2	kW/m^2	kW/m^2	kW/m^2	kW/m^2	kW/m^2	kW/m^2	kW/m^2	kW/m^2	kW/m^2
	-0.0045	-0.0086	-0.0288	-0.0111	-0.0050	-0.0028	-0.0428	-0.0031	-0.0150	-0.0927	0.0218	0.0095	-0.0409
Net Cooling Power Demand	kW	kW	kW	kW	kW	kW	kW	kW	kW	kW	kW	kW	kW
	-0.1165	-0.2223	-0.7465	-0.2869	-0.1292	-0.0718	-1.1094	-0.0805	-0.3880	-2.4019	0.5636	0.2467	-1.0606
Δ	0.0372	0.7637	0.2405	0.0536	0.4265	0.0259	1.7790	0.0299	0.1443	2.5702	0.1280	0.0542	1.7297
Δ % Case 1 Regular- Case 4 Non Regular	24%	77%	48%	23%	77%	57%	62%	27%	27%	52%	19%	18%	62%

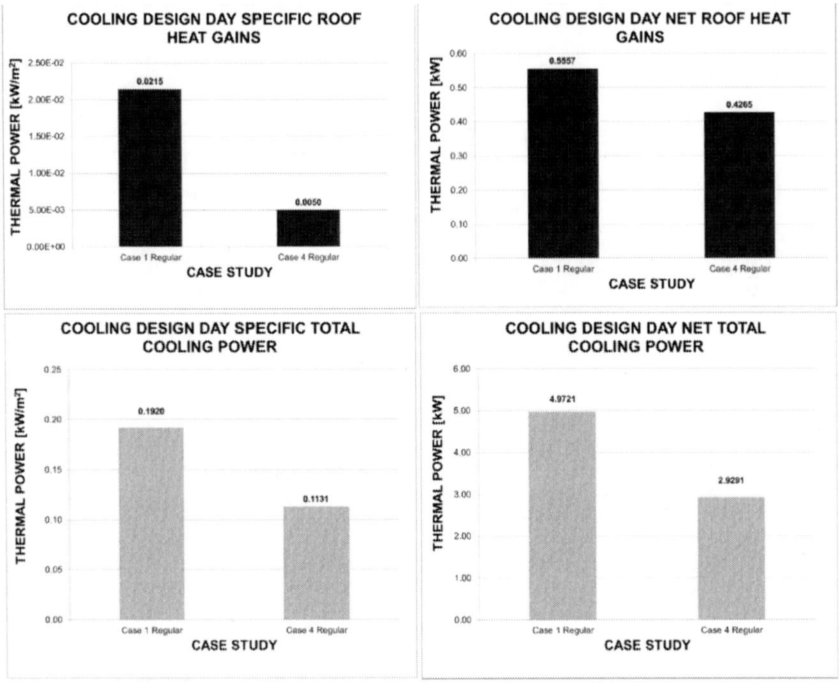

Figure 7. Regular plant distribution design day results for design day; Roof gains and cooling power.

Hourly Cooling Demand

A simulation was carried out on an hourly basis to obtain and compare the cooling demand for a typical year for case 1 regular and case 4 non-regular. The results are shown in Table 7.

As shown in Table 7 the hot water demand for absorption systems for case 4 (non-regular) is 34.05% lower than that of the base case (case 1, regular), which allows the use of fewer solar collectors and a smaller solar buffer tank. Table 8 shows the required surfaces and storage tank capacities for both case 1 (regular) and case 4 (non-regular) case studies. For the design day and the whole year, a 33.33% reduction in the initial investment is achieved. The cost of the acrylic additive for CR is approximately \$30/m^2, which is insignificant in comparison with the increase in cost for both the DX and solar-driven absorption systems. The proposed use of cool roof systems allows cost reduction for any type of HVAC system, making the initial investment more profitable. The aim of the research was to demonstrate that this scheme is profitable for both renewable and conventional driven systems. The initial investment is higher for renewable energy systems, making the use of the proposed technologies a key point for feasibility. For example, for the solar-driven system, the specific cost is approximately $\$1,050/m^2_{collector}$, which makes this option extremely profitable.

Table 7. Energy demand for a year. Results for case 1 (regular) and case 4 (non-regular)

	CASE 1. Regular distribution			CASE 4. Non-regular distribution		
	Specific total Cooling Demand	Total Cooling Demand	Total Solar Heat Demand	Specific total Cooling Demand	Total Cooling Demand	Total Solar Heat Demand
Month	kWh/m^2	kWh	kWh	kWh/m^2	kWh	kWh
	Case 1			Case 4		
1	-82	-2,125	3,572	-50	-1,305	2,194
2	-69	-1,794	3,016	-40	-1,042	1,751
3	-84	-2,168	3,645	-49	-1,281	2,153
4	-89	-2,305	3,875	-54	-1,403	2,358
5	-113	-2,932	4,927	-75	-1,935	3,252
6	-121	-3,141	5,279	-84	-2,173	3,651
7	-133	-3,433	5,769	-92	-2,386	4,011
8	-131	-3,395	5,706	-91	-2,352	3,953
9	-121	-3,139	5,276	-82	-2,128	3,576
10	-120	-3,117	5,239	-83	-2,148	3,610
11	-111	-2,883	4,845	-75	-1,949	3,276
12	-97	-2,506	4,212	-63	-1,621	2,725
Total	-1,272	-32,940	55,361	-839	-21,723	36,509
			Δ Case 1 (regular)- Case 4 (non-regular)	433	11,216	18,851
			Δ % Case 1 (regular)- Case 4 (non-regular)			34.05%

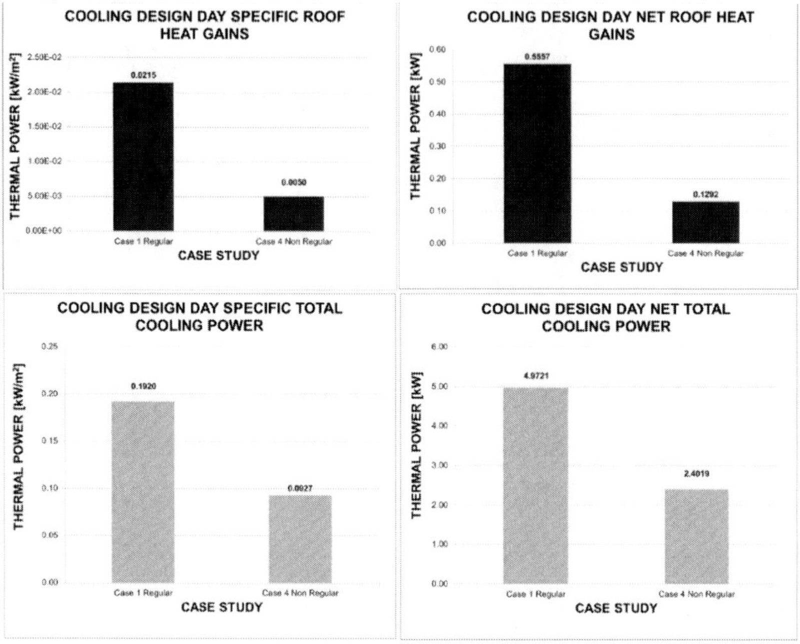

Figure 8. Non-regular plant distribution design day results for design day; Roof gains and cooling power.

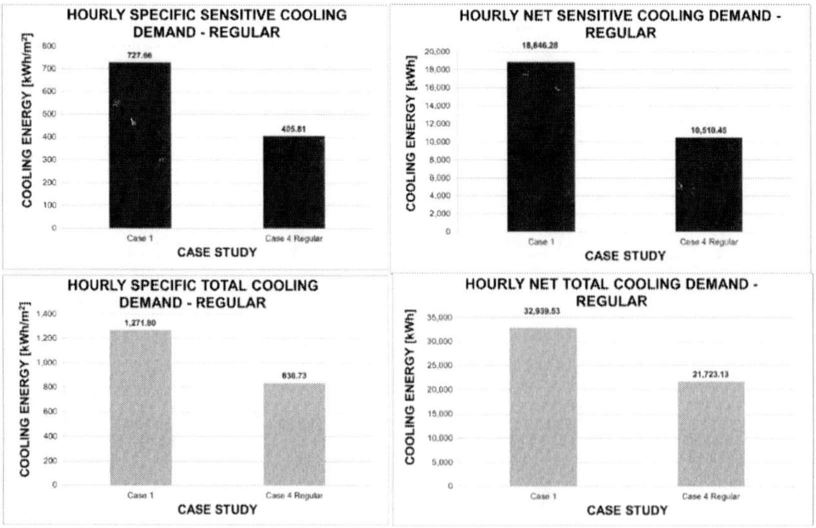

Figure 9. Regular plant distribution sensitive and total cooling demand for a year (Hourly calculation).

Table 8. Solar thermal system sizing for Case 1 (regular) and Case 4 (non-regular)

	Case 1 (regular)	Case 4 (non-regular)	Δ % Case 1 (regular) – Case 4 (non-regular)
Number of collectors	12	8	33.33%
Solar collector area (m^2)	51	34	
Solar Tank (m^3)	3,858.3	2,572.2	

Table 9. Economic analysis for case 1 (regular) and case 4 (non-regular)

	Case 1 (regular distribution)	Case 4 (non-regular distribution)	Δ Case 1 (regular)- Case 4 (non-regular)	%Δ Case 1 (regular) – Case 4 (non-regular)
Solar system initial investment cost ($)	12,600	8,400	4,200	33.33%
DX initial investment cost ($)	345	210	135	39.13%
DX yearly energy cost ($)	5,490	3,621	1,869	34.05%

For yearly operation, considering a fuel cost of $0.30/kWh, a fuel generator efficiency of 0.9, and a COP value of 2, the cost savings are presented in Table 9. The return of investment (ROI) for the cool roof is immediate for the solar-driven system and less than one year for DX systems. The low ROI makes the initial investment for passive systems affordable and increases thermal comfort while decreasing ongoing climatization costs. Using conventional DX systems for climatization can be combined with innovative energy efficiency and passive technologies to reduce both the investment and environmental impact. Moreover, a comfort level increase is achieved for inhabitants, a key point for the IDPs.

The total cooling requirements for non-regular distribution (case 1 regular and case 4 non-regular) for the whole year are shown in Figure 10.

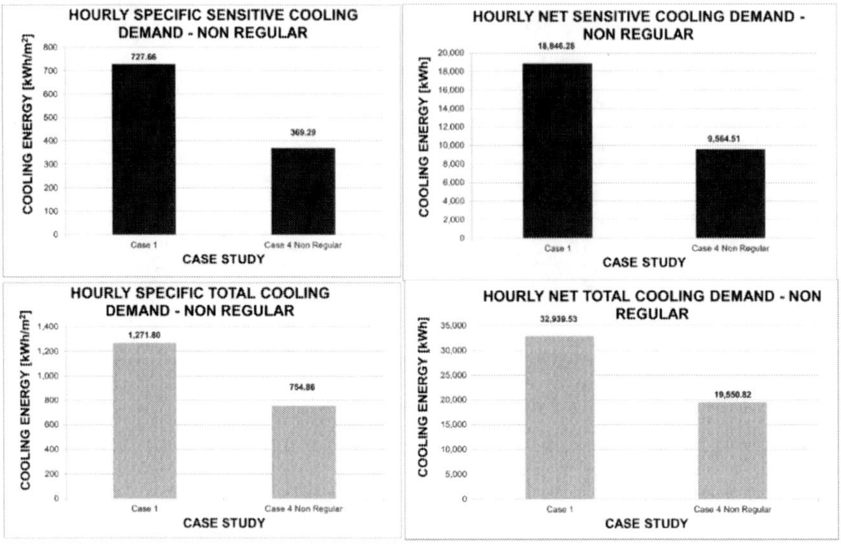

Figure 10. Non-regular plant distribution sensitive and total cooling demand for a year (Hourly calculation).

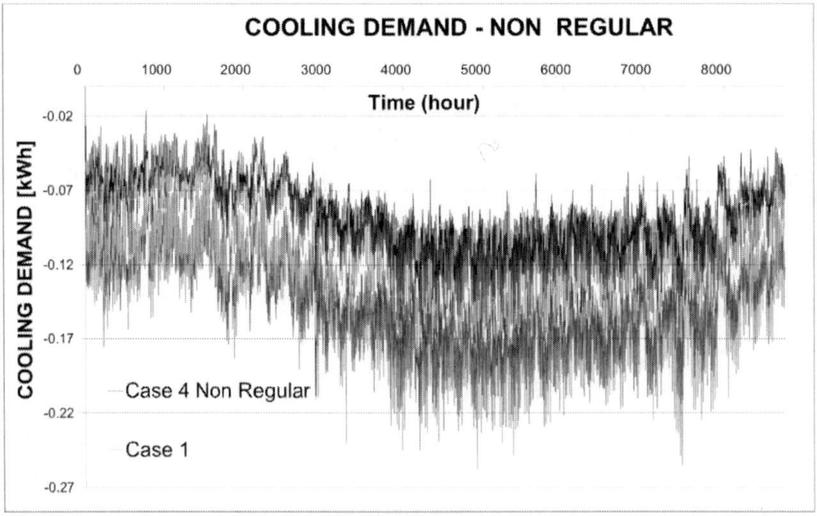

Figure 11. Hourly cooling demand results for non-regular distribution; Comparison between standard solution (case 1) and CR and Roof Opening system (case 4).

The cooling demand is up to 41% lower for case 4, with a non-regular distribution, than the base case (case 1) with a regular distribution. This result implies a reduction in fuel or electricity energy consumption for the DX system and a reduction of the initial investment cost in renewable energy systems for the solar-driven machine. The cooling demand on an hourly basis is shown in Figure 11 for a typical year for case 1 regular and case 4 non-regular. The distribution study and optimization show an extremely low-cost opportunity to improve thermal behavior in the shelters, increasing comfort and reducing initial investment and climatization costs. The external distribution is not often studied and optimized during the design phase in shelters.

Lower cooling demand is obtained for the shelter. The system both increases thermal comfort and reduces energy consumption all year, which makes the solution optimal due to a sizing reduction for the cooling design day and a reduction in energy consumption.

All the results show the capacity of passive technologies and passive design systems to improve thermal behavior with a low cost increase. Moreover, these techniques can be used in combination with both conventional and renewable energy climatization systems, reducing initial investment in both cases because of cooling load reduction. As IDPs live in very poor conditions, building optimization techniques, including no-cost improvements, such as external layout design, should be used as a general practice. CFD design and analysis tools provide a fast and reliable analysis in both the initial design phase and the building phase.

CONCLUSION

Over the last several years, building technologies that reduce energy consumption have increasingly becoming a research focus. Such systems are usually studied in the context of cities and not often applied to development cooperation. Passive techniques can be effectively used in this field to reduce energy consumption in shelters as well as the initial investment in HVAC systems. Cool roofs provide large benefits for urban use and are now being promoted all over the world. In this chapter, a combined solution is adopted by including cool roof technology, external layout distribution, and enhanced natural ventilation. The combination of these technologies provides a good solution at a low building cost. Concrete shelters are low cost and easy to build, representing a good solution for people living as IDPs or refugees. The proposed solution reduces peak power cooling demand by over 50% relative to the standard solution for the base case. The reduction in the required power implies lower investment for both conventional energy cooling systems and renewable energy HVAC equipment. When using an absorption machine, the cost reduction is for not only the cooling machine but also the solar collector system. The initial investment cost for the proposed passive techniques is minimal in comparison with the HVAC costs.

The energy required for cooling throughout an entire year is reduced by over 40%, which implies lower electricity or fuel consumption, which in turn decreases greenhouse gases emissions and improves the thermal comfort in the shelter. In the case of a solar-driven system, a lower initial investment cost is required, which makes investment more feasible. Dynamic simulation technologies allow the building design to be studied and optimized and should be of standard use for development cooperation. In buildings intended for such a use, indoor conditions are critical to provide a comfortable shelter and reduce health risks for residents.

REFERENCES

[1] Office of the United Nations High Commissioner for Refugees (UNHCR) 2010, *UNHCR*. http://www.unhcr.org, last accessed: February 2012.

[2] Borge, D., Colmenar, A., Castro, M., Sancristobal, E. Exergy efficiency analysis in buildings climatized with LiCl–H_2O solar cooling systems that use swimming pools as heat sinks, *Energy and Buildings* 43 (11) (2011) 3161-3172, http://dx.doi.org/10.1016/j.enbuild.2011.08.014.

[3] Borge, D., Colmenar, A., Pérez, C., Castro, M. Experimental validation of a fully solar-driven triple-state absorption system in small residential buildings, *Energy and Buildings* 55 (2012) 227-237, http://dx.doi.org/10.1016/j.enbuild.2012.09.040.

[4] Borge, D., Colmenar, A., Pérez, C., Castro, *M. Passive climatization using a cool roof and natural ventilation for Internally Displaced Persons in hot climates: Case study for Haiti, Building and Environment 59* (2013) 116-126, http://dx.doi.org/10.1016/j.buildenv.2012.08.013.

[5] Colmenar, L. Terán, D. Borge, M. Castro, Solutions to Reduce Energy Consumption in the Management of Large Buildings, *Energy and Buildings* 56 (2013) 66-77, http://dx.doi.org/10.1016/j.enbuild.2012.10.004.

[6] *Design Builder*, http://www.designbuilder.co.uk/, last accessed: September 2012.

[7] ASHRAE, *High Intensity Infrared Radiant Heating, 1984 system Handbook,* American Society of Heating, Refrigerating and Air Conditioning Engineers, Atlanta, 1984.

[8] ASHRAE, *Physiological Principles for Comfort and Health, 1985 Fundamental Handbook*, American Society of Heating, Refrigerating and Air Conditioning Engineers, Atlanta, 1985.

[9] Fanger, P. O. *Thermal Comfort-Analysis and Applications in Environmental Engineering*, Danish Technical Press, Copenhagen, 1970.

[10] Berglund, L. Mathematical Models for Predicting the Thermal Comfort Response of Building Occupants, *ASHRAE Transactions* 84 (1) (1978) 735—749.

[11] Du Bois, D. A Formula to Estimate Approximate Surface Area, if Height and Weight are Known, *Archives of internal Medicine* 17 (1916) 863-871.

[12] Fanger, P. O. Radiation and Discomfort, *ASHRAE Journal* 28 (2) 1986 33–34.

[13] Fanger, P. O. Calculation of Thermal Comfort: Introduction of a Basic Comfort Equation, *ASHRAE Transactions* 73 (2) (1967) III-4.1.

[14] Gagge, A. P., Stolwijk, J. A., Nishi, Y. An Effective Temperature Scale Based on a Simple Model of Human Physiological Regulatory Response, *ASHRAE Transactions* 77 (1) (1970) 247 – 262.

[15] Azer, N. Z., Hsu, S. The prediction of Thermal Sensation from Simple model of Human Physiological Regulatory Response, *ASHRAE Transactions* 83 (1) (1977) 88-102.

[16] Gagge, A. P., Fobelets, A. P., Berglund, L. G. A Standard Predictive Index of Human Response to the Thermal Environment, *ASHRAE Transactions* 92 (2) (1986) 709-731.

[17] Hsu, S. A *Thermoregulatory Model for Heat Acclimation and Some of its Application*, PhD. Dissertation, Kansas State University, 1977.

[18] Doherty, T. J., Arens, E. Evaluation of the Physiological Bases of Thermal Comfort Models, *ASHRAE Transactions* 94 (1) (1988) 1371-1385.

[19] ISO, Determination of the PMV and PPD Indices and Specification of the Conditions for Thermal Comfort, DIS 7730, *Moderate Thermal Environment*, 1983.

[20] ASHRAE, Physiological Principles and Thermal Comfort, 1993 *ASHRAE Handbook- Fundamentals*, American Society of Heating, Refrigerating and Air Conditioning Engineers, Atlanta, GA, 1983.

[21] Cooper, V. A. *Occupancy Comfort and Energy Consumption in Naturally Ventilated and Mixed Mode Office Buildings*, Ph.D. Thesis, UMIST, 1998.

[22] Fountain, M. E., Huizenga, C. A Thermal Sensation Prediction Tool for Use by the Profession, *ASHRAE Transactions* 103 (2) (1997) 130-136.

[23] de Dear, R. J. A global database of thermal comfort field experiments, *ASHRAE Transactions* 104 (1) (1998) 1141–52.

[24] Humphreys, M. A. Field studies of thermal comfort compared and applied. *Engineering Journal of Heating and Ventilation Installations* 44 (1975) 5–27.

[25] Myers, E. G. Effect of ventilation rate and board loading on formaldehyde concentration: a critical review of literature, *Forest Products Journal* 34 (10) 2004.

[26] Auliciems, Towards a psycho-physiological model of thermal perception, *International Journal Biometeorology* 25 (1981) 109–122.

[27] Baker, N. V., Standeven, M. A. A behavioural approach to thermal comfort assessment in naturally ventilated buildings, in: *Proceedings CIBSE national conference. Eastbourne,* Chartered Institute of Building Services Engineers, London, 1996, pp. 76–84.

[28] Humphreys, M. A. The influence of season and ambient temperature on human clothing behavior, in: P. O. Fanger, O. Valbjorn (Eds.), *Indoor climate*. Copenhagen: Danish Building Research, 1979.

[29] de Dear, R. J., Brager, G. S. Thermal comfort in naturally ventilated buildings: revisions to *ASHRAE Standard* 55, *Energy and Buildings* 34 (6) (2002) 549–61.

[30] Humphreys, M. A., Nicol, J. F. Understanding the adaptive approach to thermal comfort, *ASHRAE Transactions* 104 (1) (1998) 991–1004.

[31] Markov, D. *Practical evaluation of the thermal comfort parameters*, Technical University of Sofia, 2006.

[32] Olesene, B. W. A better way to predict comfort, *ASHRAE Journal* 46 (8) (2004) 20–26.

[33] Givoni, B. *Passive and low energy cooling of buildings*, Van Nostrand Reinhold, New York, 1994.

[34] Roulet, C. A. Simple and cheap air change rate measurement using CO_2 concentration decays, *Air Infiltration and Ventilation Centre technical note* (1991) 34.

[35] Nicoletti, G. M., Notarnicola, B., Tassielli, G. Comparative LCA of flooring materials: ceramic versus marble tiles, *Journal of Cleaner Production* 10 (2002) 283-296.

[36] Bolin, C. A., Smith, S. LCA of ACG-treated lumber with comparison to wood plastic composite decking, *Journal of Cleaner Production* 19 (2011) 620-629.

[37] Abeysundara, Y., Babel, S., Piantanakulchai, M. A matrix for selecting sustainable floor coverings for buildings in Sri Lanka, *Journal of Cleaner Production* 17 (2009) 231-238.

[38] EPA, *Reducing Urban Heat Islands: Compendium of Strategies Cool roofs*, 2009, http://www.epa.gov/hiri/resources/compendium.htm, last accessed: February 2012.

[39] Founda, D., Papadopoulos, K. H., Petrakis, M., Giannakopoulos, C., Good, P. Analysis of mean, maximum, and minimum temperature in Athens from 1897 to 2001 with emphasis on the last decade: trends, warm events, and cold events, *Global and Planetary Change* 44 (2004) 27–38.

[40] *LBNL Heat Island Group*, 2010, http://heatisland.lbl.gov/, last accessed: September 2012.

[41] Santamouris, M. Heat island research in Europe: the state of the art, *Advances in Building Energy Research* 1 (2007) 123–150.

[42] EPA, *Reducing Urban Heat Islands: Compendium of Strategies Urban Heat Island Basics*, 2009, http://www.epa.gov/hiri/resources/compendium.htm, last accessed: January 2012.

[43] IPCC, *Intergovernmental Panel on Climate Change*, 2007, http://www.ipcc.ch, last accessed: May 2012.

[44] Taha, H. Modeling the impacts of large-scale albedo changes on ozone air quality in the South Coast Air Basin, *Atmospheric Environment* 31 (1997) 1667–1676.

[45] Bretz, S., Akbari, H. Long-term performance of high albedo roof coatings, *Energy and Buildings* 25 (1997) 159–167.

[46] Cartalis, C., Synodinou, A., Proedrou, M., Tsangrasoulis, A., Santamouris, M. Modifications in energy demand in urban areas as a result of climate changes: an assessment for the Southeast Mediterranean region, *Energy Conversion and Management* 42 (14) (2001) 1647–1656.

[47] Hassid, S., Santamouris, M., Papanikolaou, N., Linardi, A., Klitsikas, N., Georgakis C. et al, The effect of the Athens heat island on air conditioning load, *Energy and Buildings* 32 (2000) 131–141.

[48] Santamouris, M. *Energy and Climate in the Urban Built Environment,* James and James Science Publishers, London, 2001.

[49] Konopacki, S., Akbari, H. Measured energy savings and demand reduction from a reflective roof membrane on a large retail store in Austin, Lawrence Berkley Laboratory, *Report No. LBNL-47149*, CA, 2001.

[50] Siegel, R., Howell, J. *Thermal Radiation Heat Transfer*, fourth edition, Taylor and Francis, New York, 2002.

[51] Halwatura, R., Jayasinghe, M. Influence of insulated roof slabs on air conditioned spaces in tropical climatic conditions—a life cycle cost approach, *Energy and Buildings* 41 (2009) 678–686.

[52] Konopacki, S., Akbari, H., Pomerantz, M., Gabersek, S., Gartland L. Cooling energy savings potential of light-colored roofs for residential and commercial buildings in 11 U.S. metropolitan

areas, *Paper LBNL-39433*, Lawrence Berkeley National Laboratory, Berkeley, 1997.

[53] Levinson, R., Akbari, H., Konopacki, S., Bretz, S. Inclusion of cool roofs in nonresidential title 24 prescriptive requirements, *Energy Policy* 33 (2005) 151–170.

[54] Rosenfeld, A. H., Akbari, H., Pomerantz, M., Taha, H., Romm, J. J. Policies to reduce heat islands: magnitudes of benefits and incentives to achieve them, in: *Proceedings 1996 ACEEE Summer Study on Energy Efficiency in Buildings* 9 (1996).

[55] Taha, H. Urban climates and heat islands: albedo, evapotranspiration, and anthropogenic heat, *Energy and Buildings* 27 (2007) 99–103.

[56] ASTM, E408-71, *Standard Test Methods for Total Normal Emittance of Surfaces Using Inspection-meter Techniques*, 2008.

[57] Boixo, S., Diez, M., Colmenar, A., Alonso, M. Potential energy savings from cool roofs in Spain and Andalusia, *Energy* 38 (1) (2012) 425-438.

[58] Synnefa, A. Dandou, M. Santamouris, M. Tombrou, N. Soulakellis, On the use of cool materials as a heat island mitigation strategy, *Journal of Applied Meteorology and Climatology* 47 (11) (2008) 2846-2856.

[59] Kültür, S., Türkery, N. Assessment of long term solar reflectance performance of roof coverings measured in laboratory and field, *Building and Environment* 48 (2012) 164-172.

[60] ASTM, *Standard G159-91, Standard Tables for References Solar Spectral Irradiance at Air Mass 1.5: Direct Normal and Hemispherical for a 37º Tilted Surface* (in compliance with ISO 9845-1:1992), 2003.

[61] ASTM, Standard E903-96, *Standard Test Method for Solar Absorptance, Reflectance and Transmittance of Materials Using Integrating Spheres*, 2005.

[62] Taha, H. Urban surface modification as a potential ozone air quality improvement strategy in California – phase one: initial

mesoscale modeling, Altostratus Inc. for the California Energy Commission, *PIER Energy-related Environmental Research Rep*. CEC-500-2005-128, 2005.

[63] Byrne, J., Zhou, A. M., Shen, B., Hughes, K. Evaluating the potential of small- scale renewable energy options to meet rural livelihoods needs: a GIS- and lifecycle cost-based assessment of western China's options, *Energy Policy* 35 (2007) 4391–4401.

[64] Xu, T., Sathaye, J., Akbari, H., Garg, V., Tetali, S. Quantifying the direct benefits of cool roofs in an urban setting: reduced cooling energy use and lowered greenhouse gas emissions, *Building and Environment* 48 (2012) 1-6.

[65] Alshuwaikhat, H. M., Abubakar, I. An integrated approach to achieving campus sustainability: assessment of the current campus environmental management practices, *Journal of Cleaner Production* 16 (2008) 1777-1785.

In: Alternative Energy Systems ... ISBN: 978-1-53614-203-7
Editors: A. Colmenar-Santos et al. © 2019 Nova Science Publishers, Inc.

Chapter 2

ENERGY CONSUMPTION REDUCTION IN LARGE BUILDINGS

Lya Noemí Terán de Lober, Ursula Eicker and Carlos-Ignacio Cuviella
[1]Departamento de Ingeniería Eléctrica, Electrónica, Control, Telemática y Química Aplicada a la Ingeniería, Universidad Nacional de Educación a Distancia (UNED), Madrid, Spain
[2]Hochschule für Technik Stuttgart, Stuttgart, Germany
[3]ROCA, Barcelona, Spain

ABSTRACT

The building sector consumes, even today, about one third of primary energy used in countries like Germany, which, in addition to tons of CO2 and depletion of energy sources, involves a cost of millions of Euros. This paper shows how, only through the optimization of the system technique regulations, energy savings by a margin of 30% can be achieved in offices and administrative buildings. This potential reduction or even elimination of expenses are possible with the use of the automation system that is proposed in this article. It is important to emphasize that simulation models of HVAC systems and rooms are very often used during the planning phase of a project for a proper system design. However, these simulation tools offer additional possibilities for control algorithm development and performance observation which are not used in practice so far. This article demonstrates that simulation based predictive control, using a system operation that is not depending on a main computer but on the components of an intelligent bus, offers the opportunity of a better utilization of the buildings and their thermal mass.

Keywords: adsorption coefficient of performance (COP), predictive control primary energy simulation, thermal activation of the components (TAC), thermal piles

NOMENCLATURE

BMS	Building Management System
COP	coefficient of performance
DDC	Direct Digital Control
ETFE	ethylene tetrafluoroethylene
g	energy transmission, from the German word Gesamtenergiedurchlassgrad
HVAC	Heating, Ventilating and Air Conditioning
IP	Internal Protocol
LAN	Local area Network
LON	Local Operating Network
OPC	Open Process Control
TAC	Thermal Activation of the Components
TCP	Transmission Control Protocol

INTRODUCTION

The proliferation of energy consumption and CO_2 emissions in the built environment has made energy efficiency and savings strategies a priority objective for energy policies in most countries [1].

Buildings account for 30%–40% of the total primary energy use globally [2], and the building sector offers large potential to reduce primary energy use and CO_2 emission by, e.g., reduced heating demands, increased efficiency in energy supply chains and greater use of renewable resources for materials and fuels. Several strategies can be used to realize this potential, including energy efficiency requirements in building standards, for example requirements that specify minimum energy efficiency for buildings. Improved building energy efficiency is a priority issue in the European Union (EU), where the building sector accounts for the largest share of the total primary energy use [3, 4].

The energy consumption of an office or commercial building during its operational lifetime is usually significantly higher than the energy embedded in the materials and construction. For instance, a 60-building case study of commercial buildings by Lawrence Berkeley National Laboratories showed that 50% of the buildings had control problems [5]. Analyzing the savings through operation and maintenance in 132 further buildings demonstrated that 77% of the savings were obtained by correcting control problems [6]. Energy savings in such buildings are usually in the range of 10% to 25%, and sometimes reach 44% [5]. This demonstrates the need for detailed performance observation in such buildings. However, energy management in the building sector, in most cases, still a simple collection of energy consumption data for heating cooling and electricity on a monthly or annual basis. For heating energy consumption a degree-day normalization method (e.g., German standard VDI 2067) is used to average out the influence of varying weather conditions and to allow a better comparison of the heating energy consumption of different years

[7]. For electricity consumption and cooling energy consumption such a simple normalization method does not exist. Furthermore, this backward looking method only allows the detection of variations in energy consumption over quite a long time period and the reasons for such variations in energy consumption is very difficult or nearly impossible to detect.

For a more dynamic energy management during recent years attempts were made to integrate the collection of energy consumption data into the Building Management Systems (BMS). BMS are most common in large buildings and its historic core function is to manage the environment within the building. The BMS is typically responsible for the control of heating and cooling systems and air handling units that distribute air throughout the building, and local control of the heating and cooling mixture to achieve the desired room temperature. Sometimes a secondary function is to monitor the level of human-generated CO_2, mixing outside air with waste air to increase the amount of oxygen while also minimizing heat/cooling losses. Systems linked to a BMS typically represent 40% of a building's energy usage and if lighting is included, this number approaches 70%. The BMS is therefore a critical component to managing energy demand. An improperly configured BMS with control errors can easily increase the energy consumption of buildings in the region of 5%-30% - in some cases even more [8, 9].

The main bottlenecks for an efficient energy management within building management systems are still the poor possibilities for graphical performance visualization and the cumbersome handling and analysis of measured performance data. Furthermore, the external use and exchange of data measured and collected within the local BMS is still difficult and in some cases nearly impossible.

As part of this project, the Technology Centre of Festo (hereinafter referred to as Festo TC) is analyzed in detail. Figure 1 shows the company's headquarters in Esslingen Berkheim, Germany. The core topics of the research are energy distribution, TAC, the concept of artificial light, including daylighting systems of the facade and the system technique for HVAC. The building was created with the main aim of reducing energy consumption under the given architecture and to limit design requirements to a minimum. This should achieve a minimized absolute consumption, and secondly, an overall consumption with the least possible amount of primary energy. Both approaches were implemented by selecting the appropriate components and equipment in the engineering design and building as well as through the use of an intelligent multifunctional automation system. The project focuses on building management optimization, based on the simulation with the aim to reduce the energy used for the HVAC system. Here several methods are presented, which use equipment and building simulation of the used energy. Through these methods building systems consumption can be reduced by more than 30%.

Figure 1. Company's headquarters in Esslingen Berkheim, Germany.

Building, Energy Concept, and Automation

Buildings with innovative technologies that results in very low operation energy use are being developed and may constitute a critical part of efforts to reduce fossil fuels use and CO_2 emissions [10, 11].

Construction of the Building

The modern comfort living conditions are achieved at the cost of vast energy resources. Global warming and ozone depletion and the escalating costs of fossil fuels over the last few years, have forced governments and engineers to re-examine the whole approach to the design and control of building energy system [12]. Consequently, it is of great importance in the building field to reconsider the building structure and exploit renewable energy systems, which can minimize the energy expenditure and improve thermal comfort [13].

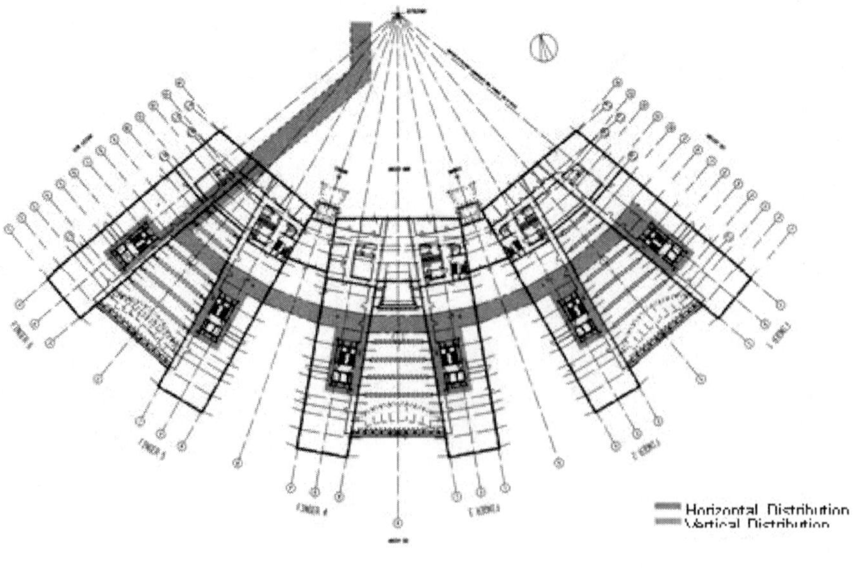

Figure 2. Plan of Technology Centre ground floor and basement.

The Technology Centre building is divided into three V-shaped building (Figure 2). Each V has two "fingers" and the six "fingers" of the building are connected on the north end. The two main entries are also located on the north side, each between the central V and an outer V. Each V has an atrium between its fingers, which has a roof and a slightly sloping facade. The spaces between the Vs are free.

Each one of the six fingers has a technical room in the basement and are connected together underground through the service tunnel. In this tunnel are located the common systems, HVAC, electricity and air pressure, and it is also a pedestrian connection between the 6 fingers and the main building of the company. The technical trenches (large pipes) are near the stairs. The foundation of the building includes approximately 380 energy piles. The whole building, except the atriums, has false floor providing a space 60 cm in height for all the technical equipment of the building [14, 15].

Building Envelop, Building Facade and Roof Construction

The exterior facade of the office building is entirely made of a high quality glass (3 layer thermal insulation glass with a high value of thermal insulation and energy transmission, which is denoted with the letter g, from the German name: Gesamtenergiedurchlassgrad). Emphasis was placed on the high air tightness of the facade. The exterior facade of the offices has external blind for sun protection. The insulation of the roof rests on a metal layer, and under this layer there is a cavity for the TAC. The facade of the three atriums has a double layer glass. Each has a shading system installed. The roof of the atriums is formed by padded sheets and the arc is mounted on a support structure. The padded sheets are under pressure and can be pneumatically conditioned to either shade in or not.

Physical Values of the Construction of the Office Area and Atriums

Offices Exterior Facade

- Overall heat transfer coefficient of the glass: U = 0.7 W/m^2 K
- Overall heat transfer coefficient of the facade (glass + construction): U = 0.8 W/m^2 K
- Total energy transmission with the opened external blind: g = 60%
- Total energy transmission with the closed external blind: g = 10%

Offices Area Roof

- Overall heat transfer coefficient of the roof: U = 0.19 W/m^2 K

Offices Area Floor

- Overall heat transfer coefficient of the floor: U = 0.24 W/m^2 K

Atriums External Façade

- Overall heat transfer coefficient of the facade (glass + construction): U = 1.45 W/m^2 K
- Total energy transmission with opened sun protection: g = 46%
- Total energy transmission with closed sun protection: g = 20%

Atriums

- Overall heat transfer coefficient of the facade (pad + construction): U = 2.7 W/m^2 K
- Total energy transmission with opened sunscreen: g = 32%
- Total energy transmission with closed sunscreen: g = 15%

Building Techniques

Sun Protection for the Offices Facade

Intelligent blinds are used with light management function included, in order to control the amount of solar radiation (light and heat) into the building. A third part of the blind (upper third) is used to control the light, and the other two thirds are used to prevent excess light and heat [16].

Sun Protection for the Atriums Facade

Sails are used; they operate with hydraulic rods, which are controlled by photosensors [16].

Sun Protection for the Atriums Roof

Roofs of inflated sheets, controlled by photosensors, are used. The two positions (open/closed) are shown in Figures 3 and 4. They consist of three layers of ETFE foil, the upper and middle layers are printed with a chessboard. The two patterns are printed so that they complement each other. If the middle layer ventilates upward shading occurs, otherwise the layer ventilates downward allowing the light to go through [17].

Figure 3. Atriums roof, open (left) closed (right).

Figure 4. Operating principle of the atriums roof, open and closed.

Production and Distribution of Heat and Cold

The building does not have its own plant to generate heat, but the heating energy is supplied from the main building. Heating water for all consumers is distributed through the technology centers of the fingers. The required refrigeration is produced by three adsorption refrigeration machines. For cooling of the refrigeration machines three cooling towers are installed. When the outside temperature is adequate, the cooling towers can be operated in free-cooling, e.g., the produced cooling is directly supplied to the consumers [18]. Furthermore, the energy piles are used in the cold supply. The distribution of heating and cooling is performed by three supply distributors. Every two fingers is supply by one of the distributors. Means for heating, cooling and the free-cooling of the building are parallel to this supply and depending on the seasons and the operating conditions of the building controlled by the control system of the building. The energy piles also provide energy in this distribution. From the supply distributors are supplied thermal activation of both heating and cooling as shown in Figure 5.

The heating and cooling of the office areas is mainly due to the TAC. The TAC is installed in the ceiling of all levels, and the walls of the staircases. Inserting the tubes of the thermal activation is shown in Figure 6. The individual loops of the TAC within a finger are connected together and share in common the temperature of the supplied flow. The various offices and meeting rooms are equipped with double bottom floor, where the equipment of refrigeration and heating is installed, in order to allow and individual climate control in every office. A system of heating and cooling by underfloor heating is used in the atriums [19, 20].

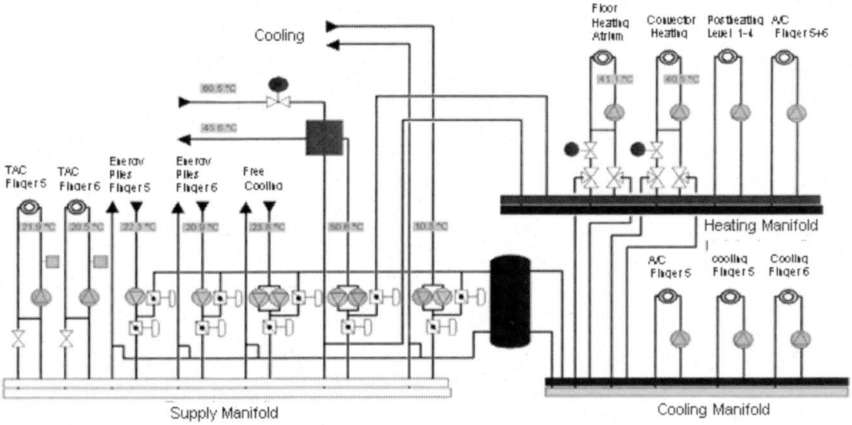

Figure 5. Heating and cooling distribution.

Figure 6. Insertion of flexible hose for TAC.

Ventilation

Each finger has its own HVAC system. On the one hand, the system provides fresh air to common office areas, the individual offices and meeting rooms. On the other hand, it compensates the differences of temperature caused by different internal and external loads. The movement of both, the incoming and outgoing air, is done through the double floor with the help from a ventilation system. The system works only with fresh air and has a high efficiency heat recovery through a rotary heat exchanger, heating and cooling coils, via a variable-frequency drive drives the direct vent and a high-pressure humidification. To compensate the temperature difference in the office area, each floor in each finger is divided into two zones, so the whole finger is divided into eight zones. The inlet air temperature in each zone can be regulated with a heater or refrigerator at the required temperature in each area. Figure 7 shows the schematic representation of the ventilation system of a finger. There is a HVAC system installed for the supply of fresh air for each atrium, which is equipped with a rotary heat exchanger of high efficiency, heating and cooling coils. The air supply is introduced through the apertures in the floor in the facade in the south side, the air extraction is carried out within the areas of galleries through the air intakes of the roof in every floor [21, 22].

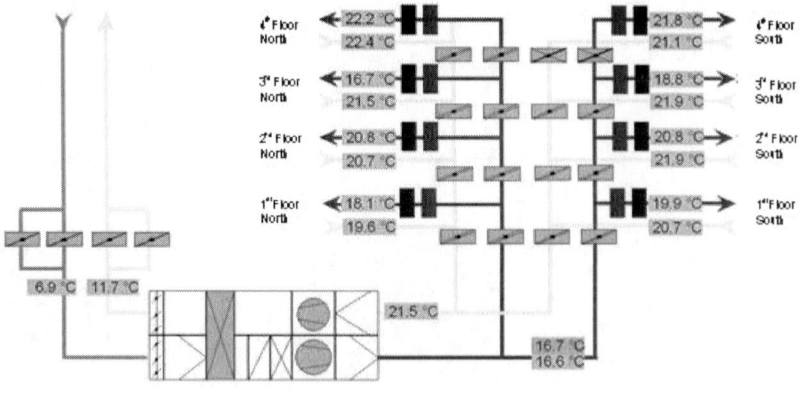

Figure 7. One finger's ventilation diagram.

Lighting

The lighting in the office areas is carried out with auto dimmer lights (automatic control system that acts depending on the natural light conditions), each light is equipped with a sensor that is directly connected to the electronic ballast. The sensor monitors the light, depending on the ambient brightness, to an adjusted light intensity. Energy is saved with this automatic control whenever the natural light conditions do not require the aid of artificial light to achieve the desired intensity. The general lighting in the large office areas are fixed in groups of 4 intercepted lights. The lighting of these groups is switched on or off manually on a touch panel for employees as needed. Approximately, 20% of the lights are assigned to the passage lights. These lights are controlled through a time switching program. The lights are switched on in the morning, at a definite time and within a time frame, when the brightness is higher than a predefined parameter, they are switched off. By using the time frame the lights are prevented from being switched on or off during the day by a higher or lower brightness. Both, passage lights like all general lights, in the whole building will be turned off at night after 8pm by a central timer program [23].

Energy Concept

The building was created with the goal of reducing the energy consumption, in the given architecture and design requirements, to limit it to the minimum. This should get to reduce the absolute consumption to the minimum and secondly the fact that the consumption is covered with the least amount of primary energy. Both approaches have been carried out by selecting appropriate components and equipment in the construction engineering, and a multifunctional building automation system.

Energy Demand Reduction

The heat energy requirements of the building are reduced by:

1. Maximum use of free solar energy during the hot season through a glass with a solar factor of 60%.
2. Reduction of heat transfer loss through the glass. Value of heat transfer coefficient U= 0,.7 W/m^2 K.
3. Reduction of ventilation heat loss through:
   - The waterproof envelope of the building
   - Fresh air supply, in accordance with specific requirements of hygiene DIN 1946 (room ventilation techniques) through a heat recovery system with a thermal efficiency of 85% recovery [24].

The cooling energy requirements of the building are reduced by:

1. The installation of an intelligent outdoor sun protection with Azimut-dependent slat adjustment and the annual shading diagram, which is integrated into the building automation system.
2. The installation of an auto dimmer lighting depends on room light [17].

Minimization of Primary Energy Demand

Heating Energy

The necessary heat is provided by three thermal production facilities:

1. Three Gas/Oil burner in the boiler room of the main building, the maximum total heating power is 4,560 kW

2. Heat recovery system of air compressors, the maximum total heating power is approximately 700 kW
3. Solar thermal power plant with a gross collector area of 1,330 m² on the roof of the main building, the peak power approximately 1,200 kW, the maximum continuous power is approximately 650 kW

For heat generation, heat recovery and solar thermal device have priority over the heating system, e.g., burners provide heat only when the solar heating and heat recovery are not sufficient to supply all consumers [25].

Cooling Energy

The cold required in the building is provided by three adsorption cooling machine with a total cooling capacity produced of approximately 1,000 kW. For the re-cooling of the cooling machines there are three cooling towers with a maximum cooling capacity installed of approximately 3,000 kW. Depending on the outside temperature and the cooling demand, the cooling towers can be used as free-cooling, e.g., the cold of the cooling towers passes directly to consumers. In addition, approximately 380 energy piles supplement with a cooling capacity of about 210 kW directly to the thermal activation of the components. The adsorption cooling machines, according to the characteristics of the system, have a maximum efficiency of 0.6, e.g., the machines require almost twice the heating energy (input energy) to produce cooling energy (output energy). Therefore, the system of adsorption cooling machines is used only when heating from heat recovery and thermal solar energy are sufficiently available. Depending on the amount of thermal energy generated, one, two or three adsorption cooling machines are running. When cold consumers require more cooling energy than produced by the adsorption cooling machines, the difference is taken from the cooling system of the main building. Through this operation, except in special situation, the system ensures that the adsorption cooling machines do not operate with primary energy [25].

Building Automation System

Basic Structure

In the Technology Centre a building automation system is installed, based on LonWorks technology (LON). The LON system is decentralized, which means that the operation of the system does not depend on a host computer, but the smart bus components communicate directly with each other. This limits the data traffic on the sections in the bus system and maximizes the reliability of the system. Through the fully standardized communication protocol and defined application profile of the application areas a wider interoperability of the various bus nodes is guaranteed. Considering the physical and logical properties, the system helps to extend or change the network. On the basis of a uniform automation system is carried out a cross-functional integration of all engineering areas of the building such as HVAC, plumbing, electricity, lighting and sun protection [26, 27]. A central computer is needed only for the maintenance or modification of the system, for example, the implementation of the new components of the bus or system, as well as for the building management system (BMS) is required. The BMS is at the management level and allows a convenient way to display both, configuration and alarm, and the implementation of higher level functions.

The physical structure of the systems consist essentially of two horizontally separate levels, the LON network (FTT-10-Bus) and a LAN TCP/IP. In the FTT-10-Bus, bus components, such as DDC, actuators, sensors, meters, etc., are connected to the LON network, LAN is the fast backbone for a long distance bridging and transmission of data to BMS computer. At the interfaces between FFT-10-Bus and LAN meets the LAN routes between the two levels. A simplified diagram of functions of the automation system of the building is shown in Figure 8.

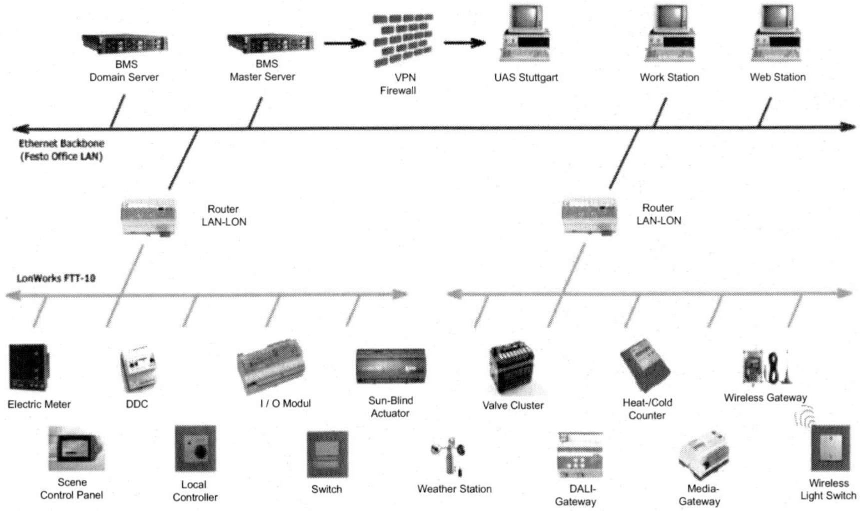

Figure 8. Schematic diagram of the system of building automation.

Control Functions

The command and control functions are, especially when they are of complex nature, implemented in the free programmable controllers (DDC), are in part firmly stored in the components of the bus. For example, controls of ventilation or heating and cooling systems in the distribution program of the DDC. The system function of sun protection, lighting or the meters of the amount of energy, however, they will be in the form of configurable standard applications, which are supplied by the manufacturer, actuators, sensors, and other bus nodes are loaded into standard applications. In order to put a system to run, all the necessary components are connected to each other through the called variable network. For example, the shutter actuators can work connected to the station of weather condition and time, to get the real values of lighting, wind power and time [18, 28].

METHODS

Buildings demand energy in their life cycle, both directly and indirectly. Directly for their construction, operation (operating energy), rehabilitation and eventually demolition; indirectly through the production of the materials they are made of and the materials technical installations are made of (embodied energy) [29]. With the evaluation of the life cycle of an industrial and administrative building, as shown in Figure 9, the operation of the building requires a large part of the total building cost. Investments in efficient energy management during the planning phase are amortized quickly and paid off during the total period of the building's life [15, 25].

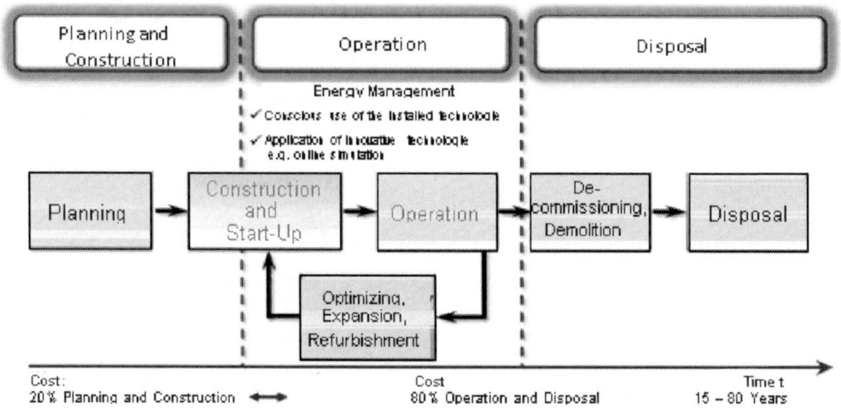

Figure 9. Life cycle of an industrial and administrative building.

While the development of the components of the building's automation in terms of increased energy efficiency is very advanced, intelligent management of the integrated systems for the optimal interaction of the components is still missing. Simulation systems offer solutions in this context, such as online tools used to monitor the automated system, for the optimized management and also to clearly reduce the necessary time for performance optimization.

Simulation Based Control Strategy Optimization

Simulation models of HVAC systems and rooms, e.g., TRNSYS and INSEL, are very often used during the planning phase of a project for a proper system design, Figure 10. However, these simulation tools offer additional possibilities for control algorithm development and performance observation which are not used in practice so far. The development of complex control algorithms of systems in the simulation environment (e.g., HVAC) offers the possibility to test and optimize the developed control code for all possible boundary conditions in a very efficient manner [26].

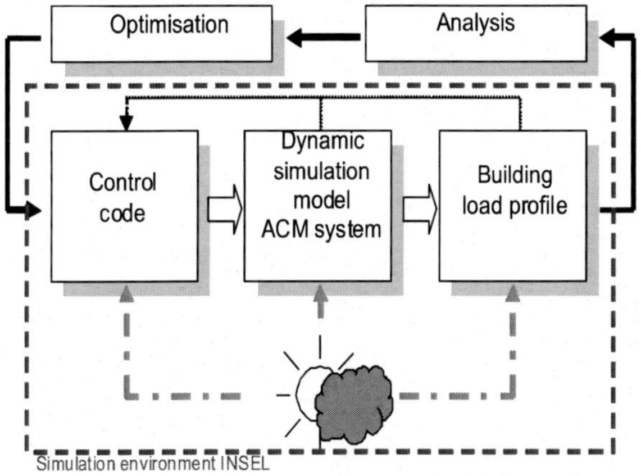

Figure 10. Principle of simulation based control optimization.

Simulation Based Performance Observation

For energy efficient operation, especially of larger and complex buildings, a permanent performance observation of the relevant HVAC systems and of the building's overall energy consumption is extremely important. The experience shows, that even systems which have once been analyzed in detail and were optimized can significantly change their operational behavior due to component degradation, system errors and false operation caused by inappropriate maintenance (valves opened and forgotten to be close, blocked water supply, wrong set points etc.). Owing to the large number of HVAC systems in larger buildings, errors of single systems are often not detected although they are monitored in the BMS. The main problem is that a short daily look at simple line graphs is often not sufficient for fault detection and time for more detailed analyses is not affordable. With the support of simulation based performance observation, the effort for a detailed performance analysis is significantly reduced, since a closer look at the performance data is only required for plants with detected errors. In the best case, the possible error source can also be detected or at least the number of possible error sources can be isolated by the simulation model [26, 27]. Figure 11 demonstrates the application of simulation based performance observation.

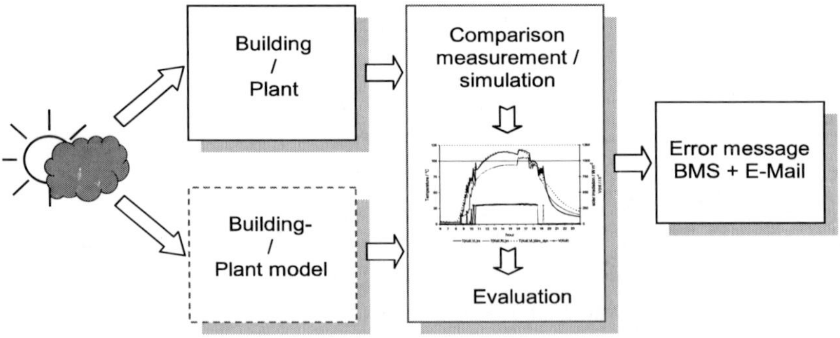

Figure 11. Principle of simulation based performance observation.

Performance Optimization through Predictive Control

The thermal mass of the building can be used as passive heat or cold storage in order to provide cooling or heating energy during periods with low primary energy effort and to reduce peak loads during periods with lower efficiency of the supply systems [27, 28]. A typical application is the utilization of night time ventilation to reduce cooling loads during the following day. In addition to ventilation, other active systems like cooling towers or chillers can also be operated during night time or early morning hours at high COP to precool the rooms and their surrounding construction to a certain temperature level [29]. Many different analyses have been performed for various building typologies, including large office buildings [30, 31]. This significantly reduces the typical peak cooling load during the early afternoon of the following day. However, care has to be taken to avoid the active or passive cooling of rooms during night-time when it is expected that the ambient conditions suddenly change and no cooling energy would be required during afternoon of the following day [32, 33]. To avoid such kinds of unsatisfying control, predictive simulation tools can be used which consider the weather prediction from a web service to analyze the most efficient control strategy of the HVAC system for the following days [35-37]. Figure 12 shows the principles of a predictive control application.

An example for predictive control we can find in the Festo TC, as it is shown in Figure 13. Comparing different control strategies to control the temperature of the TAC. In finger 2, we find a conventional control on the base of 24 hours average of the outside temperature. In finger 4, there is a predictive control that adds the temperature of the weather forecast of the next 24 hours to the last 24 hours outside temperature average. In Figure 13 we can see the outside temperature, the average of the last 24 hours and the predicted temperature calculated (the last 24 hours + the next 24 hours). Between the 16th April to the 8th of May 2009 the influence of the forecast value is clearly visible as the outside temperature fluctuations are big. Between the 9th of May and the 20th of May 2009 the influence of the forecast value is not as clearly visible as changes of the outside temperature are small.

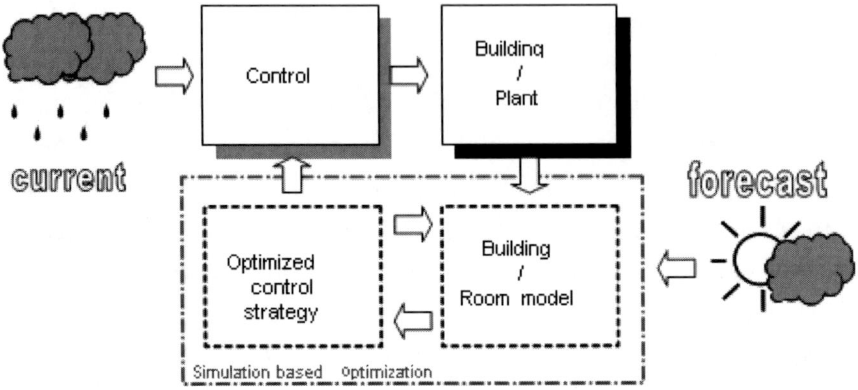

Figure 12. Principle of simulation based predictive control.

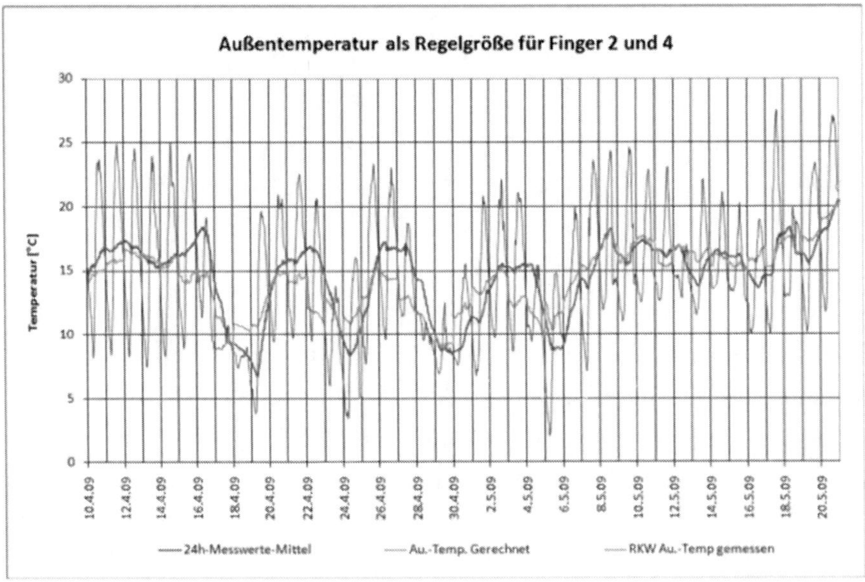

Figure 13. Measured outdoor temperature and the associated curves of the 24-hour average value, and the calculated prediction mean.

The TAC is a relatively slow system to cover the basic heating and cooling loads in the building, the control of the temperature through the calculated forecast value adjusts the power of the TAC early in time to accommodate fluctuating outside temperatures. The result of the predictive control is first a higher comfort as the room temperature fluctuates less, second the reduction of heating or cooling energy for the ventilation systems, which would have to supply the energy difference, third available energy can be used in advance through the use of the thermal mass.

Data Transfer between BMS and Simulation Tools

For the implementation of online simulation tools, interfaces are required which enable an online data exchange of measured and simulated values between the BMS and the simulation tool. From the experience in other projects this can be a very difficult and time consuming task. In the ideal case the BMS offers an OPC interface for online data exchange. In this case an OPC client of the BMS writes the actual measured values or status information etc. to an own or third party OPC server. A second OPC client of the simulation tool connects to the server and reads the required values from the OPC server, performs the simulation, and writes the results (outlet temperatures, energy consumption and error status etc.) back to the OPC server. An OPC reader client of the BMS reads the simulation results from the OPC server and transfers them to virtual data points within the BMS. These virtual data points are stored in the database and can be displayed in trend graphs together with the measured values of the real system. If a TCP/IP connection over the local network exists, the simulation tool can run on a separate PC from the one of the BMS software. Unfortunately, some of the OPC clients and servers are not clearly defined which means that they sometimes talk a kind of different ‘accent.’ This may cause serious communication problems which are sometimes difficult to solve. Figure 14 shows a standard connection solution.

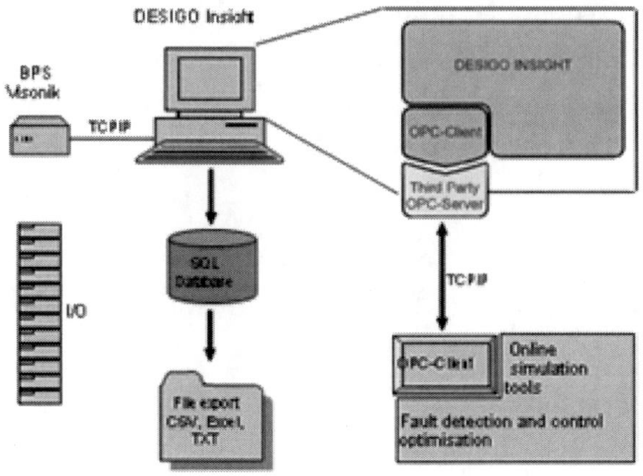

Figure 14. Simplified data communication structure for online simulations.

DETECTION OF SOME SPECIFIC ERRORS AND SOLUTIONS: OPTIMIZATION

Error Detection through Improved Performance Visualization Methods

Figure 15 shows carpet plots of the heating and cooling coil of one of the office air handling units installed in the FESTO office building. These air handling units are operated with reduced air flow rate during night time and constant design air flow rate during daytime. As clearly visible from the carpet plots the heating coil supplies heating energy to the supply air during night time and the early morning hours although the ambient temperatures (outside temperature) are not below 16°C. During daytime the ambient air temperature increases above 27°C and additional cooling energy needs to be supplied to the rooms. This inefficient behavior was caused by an error in the control of the heating coils. With the help of the carpet plots, such control errors are very fast to detect with only one short glimpse on the graphs. With simple line graphs, as still common in BMS systems, the detection of such an error would have taken much more time or even would have never been detected at all. Altogether, the energy saving potential of the detected error was more than 30% of the heating energy and a not clearly specifiable percentage of the additional cooling energy supplied to the rooms. This clearly demonstrates that for efficient energy management, improved methods of graphical presentation of performance data are essential [14].

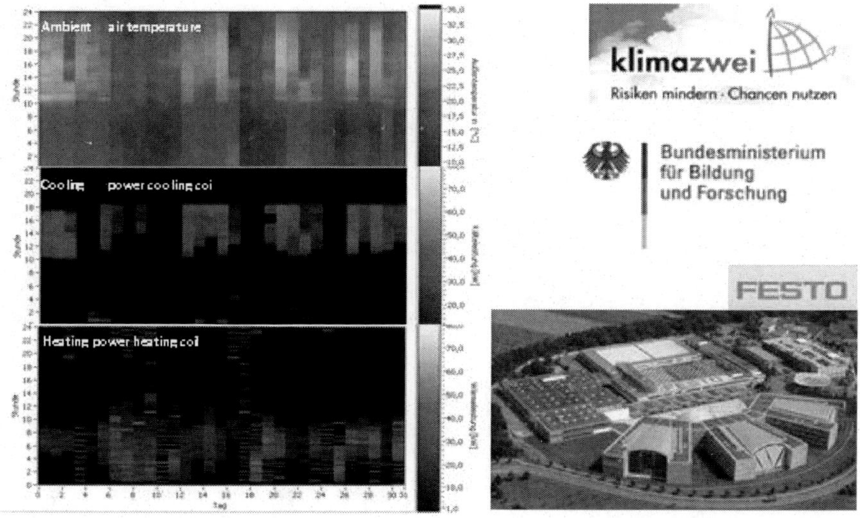

Figure 15. Carpet plots of the cooling and heating power of one air handling unit in July 2009.

Error Detection through Simulation Based Performance Observation

The main part of the cooling energy in the FESTO office building is removed by the thermally activated concrete ceilings 'TAC' which are regeneratively cooled by a large number of thermally activated bore piles in the foundation of the building. Due to a significant ground water flow rate, this geothermal cooling system is very efficient. The remaining part of the cooling energy is removed by the ventilation system (1.6 air changes per hour) and additional cooling coils. For one of the six office fingers shown in Figure 8, detailed monitoring of the energy flow dynamic simulation model have been developed for each of the 4 floors. These models include the TAC, the ventilation system, the additional cooling/heating coils and the control of the active systems. Comparative online simulations showed a very good agreement between the measured and simulated room temperatures for the ground floor, 1st floor and 2nd floor. For the 3rd floor, shown in Figure 16, a significant temperature difference of about 2 K lower was calculated by the simulation tool, which was determined to be a system error. Further analyses showed that if the TAC was turned OFF in the

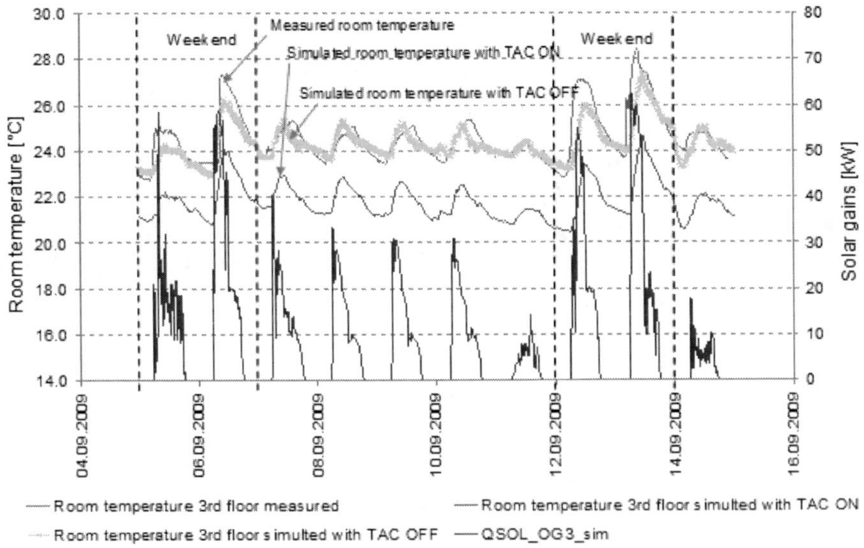

Figure 16. Measured and simulated room air temperatures for the 3rd floor of one of office finger.

simulation model nearly the same temperatures were predicted as measured in the rooms. In consequence the stopcocks of the thermally activated ceiling in the third floor were checked, with the result that all of the stopcocks were found to be closed and therefore the thermally activated ceiling wasn't active in this floor at all. In consequence the part of conventional cooling energy was much higher. This example clearly demonstrates the potential of simulation tools for automated fault detection and fault diagnosis. Without the simulation tool the detection and localization of the error would have been much more difficult and time consuming.

Simulation Based Control Strategy Optimization

As already described in the previous chapter, the main cooling load in the FESTO building is removed by thermally activated ceilings which are connected to a regenerative geothermal cooling system. The goal of the implemented control is to remove as much cooling energy as possible by this system. Therefore, in the actual control, the pump of the thermally activated ceilings is permanently in operation. The supply temperature is controlled in dependence on the external temperatures. In winter mode heating energy is supplied to the system by a large 1 218 m^2 solar collector field and, if this is not sufficient, additionally by gas boilers. However, a detailed analysis of the performance data showed a significant heating energy demand during the cooling period, which indicates that due to the permanent operation of the thermally activate ceilings sometimes too much cooling energy is supplied to the rooms. Case 0 in Figure 17 shows the primary energy consumption for heating and cooling and the actual control strategy for the period from March to end of September 2009 for one of the six office fingers of the FESTO office building in Esslingen. For the cooling energy delivered by the thermally activated ceiling, the primary energy consumption was calculated from the electricity consumption of the pump multiplied by the primary energy factor for electricity (this factor is used to calculate the quantity of primary energy necessary to produce electricity). According to GEMIS [38], a program that is used for the analysis and database of energy systems, materials and transport, a factor of 2.7 can be used for the electricity mixture in Germany, e.g., in order to produce 1 kWh of electricity in Germany 2.7 kW of primary energy are needed. For the additional cooling energy an electrical COP of 3 was considered for the chillers. For the additional heating a gas boiler with a primary energy factor of 1.2 was assumed. With the validated simulation model of the building and the HVAC system the effects of different operation regimes with shorter operation of the pump (12 hours or 14 hours instead of 24 hours) were analyzed. In case 1 the pump of the TAC is in operation from 0 to 12am, in case 2 from 4am to 6pm and in case 3 from 7am to 7pm.

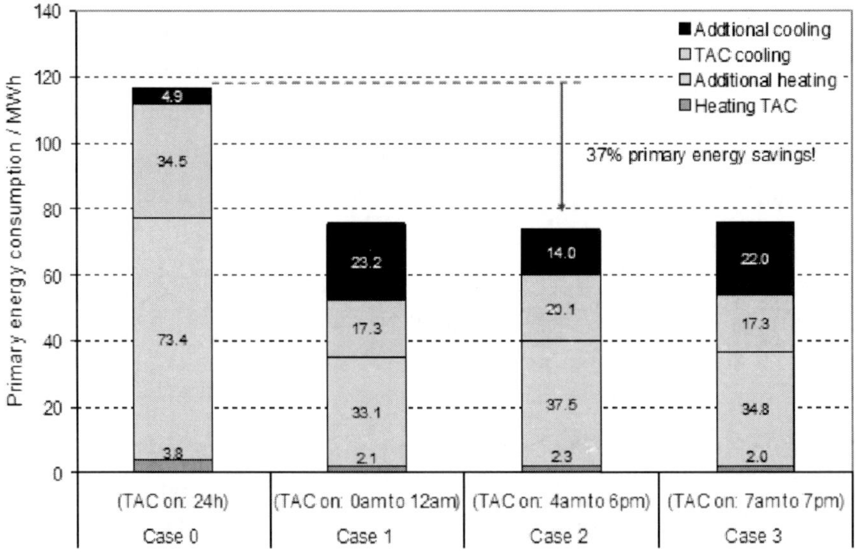

Figure 17. Primary energy consumption for heating and cooling in one office finger.

As clearly shown in Figure 10, the reduced operation time of the TAC significantly reduces the overall primary energy consumption by 37% in the best case. The reason for this is that the primary energy consumption for heating is reduced by more than 50% and the primary energy consumption of the TAC (only caused by the pump) is reduced by 50% in case 1 and 3 and by 42% in case 2. The required additional cooling energy demand increases due to the shorter operation time of the TAC by a factor of nearly 5 in case 1 and 3 and by a factor of 3 in case 2. This demonstrates that with a simple reduction of the operation time of the TAC, a significant reduction of the primary energy consumption of more than 35% is reached. However, there is still an optimization potential of 10% to 20% if the TAC operation time would be controlled dynamically using weather forecast data together with a simulation based predictive control optimization tool as described in point 2.

CONCLUSION

The simulation based control optimization methods includes control strategy development, test within the simulation environment and simulation based performance observation with automated error detection through predictive control optimizations using simulation tools together with weather forecast data.

As described and demonstrated, these tools allow a very efficient error detection of false installations and slow degradations as well as errors caused by inappropriate maintenance. Furthermore, simulation based predictive control offers the opportunity of a better utilization of the buildings and their thermal mass. This helps to equalize peak loads in the early afternoon and to operate the active cooling systems more efficiently at higher power, e.g., during the early morning hours. The expected optimization potential of the described tools is in the region of 10% to 30% or more of the actual energy consumption of the buildings, this percentage is variable depending on building characteristics and the management strategies.

REFERENCES

[1] Pérez-Lombard, J. Ortiz, C. Pout, A review on buildings energy consumption information, *Energy and Buildings* 40 (2008) 394-398.

[2] United Nations Environment Programme, *Buildings and Climate Change Status, Challenges and Opportunities*, United Nations Environment Programme, Paris, 2007, ISBN: 9789280727951.

[3] *European Commission, Action Plan for Energy Efficiency: Realising the Potential*, 2006, Web accessed at http://ec.europa.eu on June 9, 2010.

[4] Borge, D., Colmenar, A., Castro, M., Martín, S., Sancristobal, E. Exergy efficiency analysis in buildings climatized with LiCl-H_2O solar cooling systems that use swimming pools as heat sinks, *Energy and Buildings* 43 (2011) 3161-3172.

[5] Claridge, D. E., Haberl, J., Liu, M., Athar, A. `Can you achieve 150% of predicted retrofit savings?' *Proceedings of the ACEEE 1994 Summer Study on Energy Efficiency in Buildings*, American Council for an Energy Efficient Economy, Washington D.C.

[6] Hicks, T. W., Neida, B, 'An Evaluation of America's First Energy Star Buildings' *Proceedings of the ACEEE 2000 Summer Study on Energy Efficiency in Buildings*, American Council for an Energy Efficient Economy, Washington D.C.

[7] *VDI 2067/Heizgradtage*. Raumheizung: Berechnung der Kosten von Wärmeversorgungsanlagen. [*2067/heating degree days.* Space heating: Calculation of the costs of heat supply systems].

[8] Fisch, M. N., 'EnBop – Energetische Betriebsoptimierung. '*Das neue Forschungsfeld im F+E-Programm Energieoptimiertes Bauen (EnOB) des BMWi*,' ICEBO`08, 20.-22., October 2008, Berlin, Germany. ['*EnBop - Energetic Operational Optimization. 'The new research field in the R + D program Energy-Optimized*

Construction (EnOB) of the BMWi,' ICEBO`08, October 20-22, 2008, Berlin, Germany].

[9] Neumann, C., *'ModBen – Methoden zur modellbasierten Betriebsanalyse von Bestandsgebäuden,'* Workshop 'Kriterien und Werkzeuge für ein kosteneffizientes Energiemonitoring' in the frame of ICEBO`08, 22. October 2008, Berlin, Germany. [*'ModBen - Methods for model-based operational analysis of existing buildings,'* Workshop ''Criteria and tools for cost-efficient energy monitoring' in the frame of ICEBO`08, 22 October 2008, Berlin, Germany].

[10] Dodoo, A., Gustavsson, L., Sathre, R. Building energy efficiency standards in a life cycle primary energy perspective, *Energy and Buildings* 43 (2011) 1589-1597.

[11] Yu, Z., Fung, B. C. M., Haghighat, F., Yoshino, H., Morofsky, E. A Systematic procedure to study the influence of occupant behavior on building energy consumption, *Energy and Buildings* 43 (2011) 1409-1417.

[12] Florides, G. A., Tassou, S. A., Kalogirou, S. A., Wrobel, L. C. Review of solar and low energy cooling technologies for buildings, *Renewable and Sustainable Energy Reviews* 6 (2002) 557–572.

[13] Zhai, X. Q., Wang, R. Z., Dai, Y. J., Wu, J. Y., Xu, Y. X., Ma, Q. Solar integrated energy system for a green building, *Energy and Building* 39 (2007) 985-993.

[14] Eicker, U. und Seeberger, P. *Evaluierung Passiv-Bürohaus Lamparter* [*Evaluation passive office building Lamparter*], Stuttgart: s.n., 2005.

[15] Eicker, U., et al. *Evaluierung eines im Passivhaus-Standard sanierten Bürogebäudes hinsichlich des energetischen und raumklimatischen Verhaltens.* [*Evaluation of an office building refurbished in accordance with the Passive House standard with regard to the energetic and indoor climate behavior*]. Stuttgart: s.n., 2007.

[16] Eicker, U., Bauer, U. und Fux, V. *Optimierung von Fassaden zur Vermeidung von sommericher Überhitzung* [*Optimization of facades to avoid summer overheating*]. Stuttgart: s.n., 2005.

[17] Edo, W., Jakub, W., und Anna, R. *Betriebserfahrungen aus solar unterstützter Kaltwassererzeugung zur Gebäudeklimatisierung* [*Operating experience from solar assisted cold water production for air conditioning*]. 2009.

[18] Dalibard, A., et al. - Performance analysis and optimisation through system simulations of renewable driven adsorption chillers - *2nd International Conference solar Air-Conditioning.* Tarragona, Spain: s.n., October 2007. Seite 269.

[19] Olesen, Bjarne und Corro Dossi, Francesco. Neue Erkenntnisse über Regelung und Betrieb für die Betonkernaktivierung [New knowledge about control and operation for concrete core activation]. *Heiztechnik.* März 2005, Bd. 56, S. 35-40.

[20] Olesen, Bjarne und Curro Dossi, Francesco. Neue Erkenntnisse über Regelung und Betrieb für die Betonkernaktivierung [New knowledge about control and operation for concrete core activation]. *Heiztechnik.* Januar 2005, Bd. 56, S. 29-34.

[21] Arbeitsstätten-Richtlinie. *§6 Raumtemperaturen* [*§6 Room temperatures*]. 08. Mai 2001 (BArbBl. 6-7/2001).

[22] Huber, K. Detailmonitoring einer solarthermischen Anlage zur Unterstützung der Kälteversorgung eines Büro- und Verwaltungsgebäudes [Detailed monitoring of a solar thermal system to support the cooling of an office and administration building]. *18. Symposium Thermische Solarenergie.* Staffelstein: s.n., 2008.

[23] Voss, K. und Hoffmann, C. Zur Ermittlung des Energiebedarfs für Kunstlicht in Bürogebäuden - Diskussion von Meßresultaten [To determine the energy requirement for artificial light in office buildings - discussion of measurement results]. *Bauphysik* 27 - Heft 4-. 2005.

[24] DIN 1946, Teil 2. "*Anforderungen an Raumlufttechnische Anlagen.*" ["*Requirements for ventilation and air conditioning systems.*"]

[25] Weber, L. *Energie in Bürogebäuden – Verbrauch und Energie relevante Entscheidungen*. ETH Zürich: vdf Hochschulverlag, 2002. [*Energy in office buildings - consumption and energy relevant decisions*. ETH Zurich: vdf Hochschulverlag, 2002].

[26] Pietruschka, D., et al. Simulationsbasierte Planung solar betriebener Kälteanlagen an zwei Beispielen aus der Praxis [Simulation-based planning of solar powered refrigeration systems based on two practical examples]. *17. Symposium Thermische Solarenergie*. Staffelstein: s.n., 2007.

[27] Schumacher, J. *Digitale Simulation regenerativer elektrischer Energieversorgungs-systeme* [*Digital simulation of regenerative electrical energy supply systems*] - Dissertation Universität Oldenburg: s.n., 1991.

[28] Torrens, J., Keane, M., Costa, A., O'Donnel, J. *Multi-criteria optimization using past, real time and predictive performance benchmarks, 2009*, Department of Civil Engineering, National University of Ireland, Galway.

[29] Clarke, J. A., Cockroft, J., Conner, S., Hand, J. W., Kelly, N. J., Moore, R., O'Brien, T., Strachan, P. Simulation-assisted control in building energy management systems, *Energy and Buildings* 34 (2002) 933-940.

[30] Lee, S. U., Painter, F. L., Claridge, D. E. *Whole-building commercial HVAC system simulation for use in energy consumption fault detection*, TEES, Department of Mechanical Engineering, 2007.

[31] Zhou, D., Hong Park, S. Simulation-Assisted Management and Control Over Building Energy Efficiency – A Case Study, *Energy Procedia* 14 (2012) 592-600.

[32] Clarke, J. A., Cockroft, S., Hand, J. W., Kelly, N. J., Moore, R., O'Brien, T., Strachan, P. Control In *Building Energy Management*

Systems: The Role of Simulation, Seventh International IBPSA Conference, Rio de Janeiro, Brazil, August 13-15, 2001.

[33] Doukas, H., Patlitzianas, K. D., Iatropoulos, K., Psarras, J. Intelligent building energy management system using rule sets, *Building and Environment* 42 (2007) 3562-3569.

[34] H. Doukas, C. Nychtis, J. Psarras, Assessing energy-saving measures in building through an intelligent decision support model. *Building and Environment* 44 (2009) 290-298.

[35] Modellprädiktive Regelung für Heizungsanwendungen [Model predictive control for heating applications]. Peter Gruber, Markus Gwerder, Jürg Tödtli. *SGA- Bulletin* Nr. 34: s.n., Januar, Februar, März 2002.

[36] Sartori, I., Hestnes, A. G. Energy use in the life cycle of conventional and low-energy buildings: A review article, *Energy and Buildings* 39 (2007) 249-257.

[37] Himmler, Robert. *Methode zur wetterprognosegeführten Regelung von Systemen zur Betonkerntemperierung* [*Method for weather prognosis-controlled control of concrete core tempering systems*]. Lübeck und Marburg: Tönning, 2008. S. 104 ff.

[38] *Global Emission Model for Integrated System*s (GEMIS) Version 4.7 http://www.gemis.de/en/index.htm accessed 02/02/2012.

In: Alternative Energy Systems … ISBN: 978-1-53614-203-7
Editors: A. Colmenar-Santos et al. © 2019 Nova Science Publishers, Inc.

Chapter 3

PASSIVE CLIMATIZATION IN HOT CLIMATES

Clara Pérez-Molina*[*] *and Francisco Mur-Pérez
Departamento de Ingeniería Eléctrica, Electrónica, Control,
Telemática y Química Aplicada a la Ingeniería,
Universidad Nacional de Educación a Distancia (UNED),
Madrid, Spain

ABSTRACT

The 2010 Earthquake in Haiti caused catastrophic damage to the metropolitan area of Port au Prince. The earthquake destroyed approximately 105,000 homes, causing more than 2.3 million people to live as Internally Displaced Persons (IDPs). Today, more than 1.3 million people still live in tents. Low cost buildings can provide a housing solution for these people. This paper proposes a low cost building solution intended for hot climates, using concrete as the only material. Due to the lack of facilities at these locations, no conventional energy or cooling systems can be installed; thus, only passive cooling technologies can be used to increase thermal comfort. A low-cost cool roof and combined natural ventilation is proposed, and simulations show an

[*] Corresponding Author Email: clarapm@ieec.uned.es.

improvement of 16% in the thermal conditions inside the building. The simulation is performed using the software package EnergyPlus and shows that cool roofs can be a good solution for improving living conditions in low-cost houses for IDPs.

Keywords: low cost housing, cool roof, thermal comfort, development cooperation; international energy policy

NOMENCLATURE

ASHRAE	American Society of Heating, Refrigerating and Air-Conditioning Engineers
CFD	Computer Fluid Dynamics
CIDPs	Camps of Internally Displaced People
EPS	Expanded Polystyrene Insulation
IDPs	Internally Displaced Persons
h	Heat Transfer Coefficient
HVAC	Heating, Ventilation, Air Conditioning
M	Metabolic rate (W/m^2)
P	Pressure (Pa)
PMV	Predicted Mean Vote
PPD	Predicted Percentage Dissatisfied
RH	Relative Humidity
T	Temperature
U	Global transmittance value (W/m^2K)
W	External work (W/m^2)

Subscripts

a	Air
c	Comfort

cl	Clothing
d	Door
fra	Frame
glaz	Glazing
m	Mean outdoor temperature
op	Operative
r	Radiant
w	Partial water air pressure

INTRODUCTION

Internally Displaced Persons (IDPs) are a common problem around the world due to military conflicts, natural catastrophes and political conflicts. In many countries around the world, IDPs still live in tents or transitional shelters in very poor conditions, with a general lack of thermal comfort.

The purpose of buildings is to provide shelter and comfortable living conditions for the people inside. Most of the tents and transitional shelters used in Camps of Internally Displaced Persons (CIDPs) are characterized by very poor indoor air quality. In general, the building materials are extremely substandard and have high water absorption, high thermal conductivity and low compressive strength. In very few places, some type of energy supply is available for the CIDPs, but where it is available, the cooling and heating systems have very high energy consumption and poor thermal performance. There are two different aspects to address in this situation. The first such consideration is the building itself (for example, cracking walls), and the second consideration is related to the energy performance and thermal comfort inside, typically involving significant indoor temperature fluctuations. In these situations, the house design does not follow passive solar design features, which implies higher energy consumption, higher costs during the life of the building and low-comfort indoor conditions. As a result, there is a need to improve the thermal efficiency of these low-cost shelters or provide a cost efficient solution. When no energy supply is available in the CIPD and no renewable energy systems can be used due to economic restrictions, a passive design is the only way to optimize the thermal conditions. Simple design considerations such as northern orientation in the southern hemisphere and correct material choice can significantly increase the comfort conditions in buildings. In extreme conditions, which are typical in countries undergoing military conflicts or after a natural disaster, there is usually no choice about these issues.

One of these cases is Haiti, where the January 12th 2010 earthquake caused catastrophic damage to the Port au Prince metropolitan area and underscored the endemic weakness of the infrastructure and tenure regime of the country. In addition to existing problems, the earthquake created a host of new challenges. The earthquake destroyed approximately 105,000 homes and severely damaged more than 208,000, exacerbating the already substantial housing deficit and causing an estimated $4.3 billion in damage to physical infrastructure. The destruction displaced roughly 2.3 million Haitians, of whom 1.3 million still live in tents or transitional shelters. CIDPs are scattered across the landscape. These emergency settlements offer insecure living conditions and perpetuate social decline. The lack of available shelter options for people who have lost both their livelihood and home has impeded the return of IDPs to sustainable shelter. IDPs must return to safe, durable shelter in their communities or find sustainable shelter in new economically viable areas. Providing safe shelter for those displaced is a difficult, costly and humane challenge for the building designers. This paper aims to analyze and optimize the construction of 204 homes arranged in building lots, as shown in Figure 1. The scenario is based on a real location in Haiti that is intended for IDPs camps. The exact north location is shown in the figure.

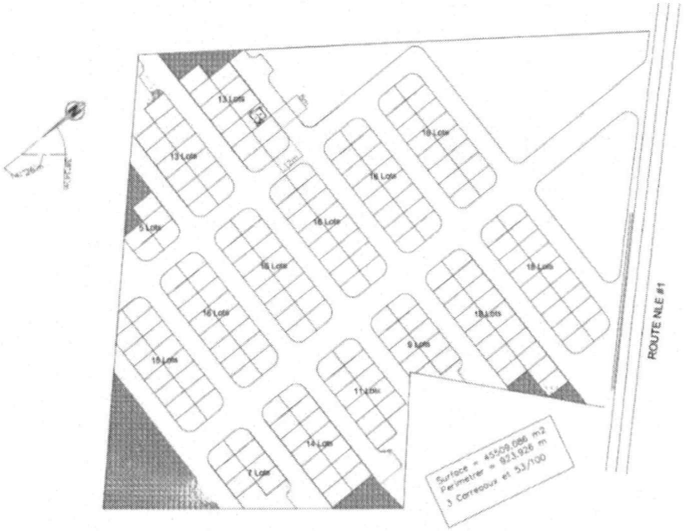

Figure 1. Lots plan.

In this paper, passive cooling techniques are studied and applied to optimize the indoor conditions. Different passive heating and cooling technologies are reported in the literature. In the hot and humid climate conditions of Haiti, passive cooling is intended for the proposed solution. Table 1 shows the main climatic conditions for the location.

Passive cooling systems in buildings can generally be studied in two different ways. The first way is to consider any system intended to reduce the cooling load as a passive cooling technique. The second way considers an integrated building design and the use of renewable energy systems to enhance heat loss from the building. Lowering the indoor temperature can be achieved using the ambient air, the upper atmosphere, water evaporation and under-surface cooling coils. Some of these cooling systems include naturally ventilated buildings, nocturnal ventilation cooling, indirect evaporative cooling, direct evaporative cooling and the use of cooled soils. In the proposed building, no active cooling systems or renewable energy systems are used. Table 2 shows the most important passive cooling techniques, their description and their energy saving potential.

Table 1. Location main climatic conditions

Month	Air temperature	Relative humidity	Daily solar radiation–horizontal	Atmospheric pressure	Wind speed	Earth temperature	Heating degree-days	Cooling degree-days
	°C	%	$kWh/m^2/d$	kPa	m/s	°C	°C-d	°C-d
January	24.6	67.4%	4.71	99.3	5.4	26.2	0	452
February	25.0	64.2%	5.17	99.3	5.3	26.9	0	420
March	25.7	62.9%	5.51	99.2	5.0	28.2	0	486
April	26.2	65.2%	5.88	99.1	4.4	29.0	0	485
May	26.3	71.1%	5.85	99.1	4.5	28.8	0	507
June	26.3	74.9%	6.22	99.2	4.9	28.2	0	489
July	26.0	75.5%	6.27	99.3	5.4	27.6	0	497
August	26.1	76.5%	6.09	99.2	5.0	27.7	0	498
September	25.9	78.0%	5.42	99.0	4.3	27.7	0	477
October	25.7	77.6%	4.86	99.0	4.1	27.5	0	486
November	25.3	74.6%	4.46	99.1	5.0	26.8	0	458
December	24.8	70.3%	4.40	99.2	5.4	26.3	0	459
Annual	25.7	71.6%	5.40	99.2	4.9	27.6	0	5,713

Table 2. Passive cooling techniques

Type of technique	Description	Costs	Savings
Operational management	Regulation of internal temperature in each space	null	medium-low
	Increase ambient temperature	null	low
	Increase ambient relative humidity	null	low
	Correct utilization of the lighting plant and electrical appliances	null	medium-low
	Correct management of external windows and shutters	null	medium
Reduction of internal thermal loads	Regulation of the lighting plant system	low	medium
	Utilization of low-energy lighting appliances	medium	high
Interventions on the building envelope	Internal shading devices	low	medium-high
	External shading devices	medium	high
	Application of vertical or horizontal o	high	high
	Application of reflective double glazing	high	high
	Application of reflective film	medium	medium-high
	Painting of external walls or roofs with light colors with low absorption	low	low
	Insulation coating of perimeter	high	medium-low
	Application of ventilated cavity wall	high	medium
	Roof insulation	medium	medium
	Application of shading devices on the roof covering	high	high
	Ventilated roofs	high	medium-high
	Garden roofs	high	high
Intervention on plant system	Installation of a heat recovery unit from extracted air	high	high
	Implementation of free-cooling and night purge	medium	medium
	Installation of efficient regulation systems	high	medium
	Installation of radiant terminals (cold ceilings, cold beams, etc.)	high	medium

Due to the low cost requirements for the building and the location, the use of a cool roof, roof insulation and natural ventilation was selected. Due to the lack of economic and financial resources, the government and the United Nations (UN) proposes the construction of low cost buildings for the IDPs that could become future stable buildings when any means of energy supply is reestablished or installed. The cost must be extremely low, and the building techniques should not be difficult because the buildings will be constructed by unqualified people with a lack of resources. Lightweight concrete buildings provide a cost-effective solution because they are easy to build and have a high thermal inertia. Ten inhabitants are intended to live in each shelter, and there will be no Heating Ventilation Air Conditioning (HVAC) systems, electricity or primary energy used. In this paper, a low cost solution for improving thermal conditions is proposed, evaluated and simulated. For the location considered, the building distribution in the lots is analyzed using an external Computer Fluid Dynamics (CFD) simulation to optimize wind distribution and ventilation. Natural ventilation for the building itself is simulated and optimized, and the use of EPS insulation in the roof is compared with the base case. A cool roof technology using a coating intended for that specific feature is proposed. Comfort parameters are evaluated for every case to obtain more detailed results than are possible using only dry or wet bulb temperatures. For a simulation combining the effects of natural ventilation and a cool roof, comfort conditions are 20% better than those attained in the base case; the operative temperature is 25.85ºC on the hottest day of the summer, while it is 28.75ºC for the base case. For the natural ventilation technique, the wind velocity, frequency and direction are important to ensure correct behavior. The wind data for the location are shown in Figure 2, and the speed frequencies are detailed in Figure 3. The most frequent direction of the wind is east, which has been considered in the distribution of the buildings in the lots.

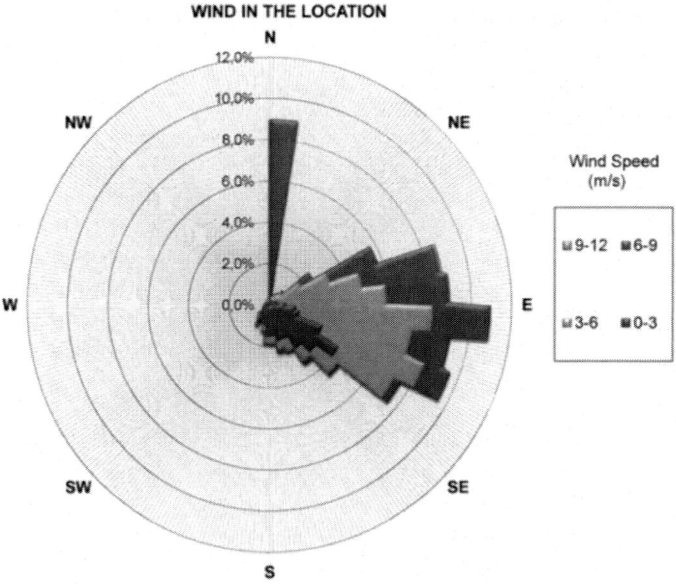

Figure 2. Wind characteristics in the location.

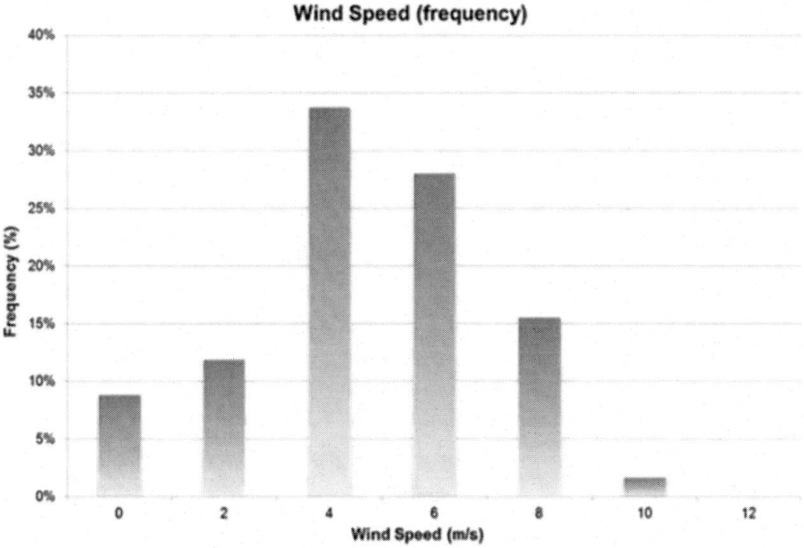

Figure 3. Wind speed and frequency.

BUILDING DESIGN

Fundamentals of Thermal Comfort

Air temperature and relative humidity have a significant impact on the perception of thermal comfort by human beings. A relative humidity (RH) below 30% may cause dry skin, eye irritation and/or respiratory problems. On the other hand, a RH above 60% may provide an environment prone to the growth of mold, mildew and dust mites and cause allergic reactions. The American Society of Heating, Refrigerating and Air-Conditioning Engineers (ASHRAE) recommends indoor comfort values [1, 2].

Human activities such as cooking, bathing, breathing, and maintaining houseplants introduce indoor pollutants. It is therefore necessary to evaluate the indoor air quality and improve the daily management of the ventilation components. The ASHRAE recommends an average natural infiltration rate of 0.35 m^3/hm^2 (air change rate) and an indoor carbon dioxide concentration of less than 0.500%.

Thermal comfort can be assessed by many parameters. Humphreys investigated the thermal neutrality of the human body and defined that parameter as the temperature at which the person feels thermally neutral or "comfortable" [4]. In his study, he used the available data from more than 30 comfort surveys around the world and proposed a series of simple correlations for thermal comfort prediction [3]. After a statistical data analysis using regressions, the results showed that 95% of the neutral temperature is associated with the variation of the outdoor mean temperature. In the case study proposed in this paper, the building has no heating or cooling systems. For this type of building, many other researchers have since found similar results [5]. A regression equation used for buildings with no HVAC is shown in Eq. 1, [4-8].

$$T_c = 0.534T_m + 11.9 \tag{1}$$

In this equation, T_c is the predicted indoor comfort temperature, and T_m is the monthly outdoor temperature. Other methods of thermal comfort evaluation were used in this study, such as the Predicted Mean Vote (PMV) method and the Predicted Percentage Dissatisfied (PPD). These parameters can be used for comfort evaluation. According to the ISO 7730 norm, the PPD should not exceed 10%. PMV and PPD provide an accurate measure of the response of people to thermal comfort. In the PMV, an index ranging from -3, which indicates a response of very cold, to +3, which indicates a response of very hot, is used. The combination of both parameters is optimal for comfort evaluation because PPD provides information in terms of the acceptability of the environment, whereas PMV shows whether the temperature is too hot or too cold, [9] A calculation of energy balances for the human body is used to obtain the PMV. Human activity, metabolic rate, clothing and thermal resistance are used in the human body model. For the environment, the air temperature, mean radiant temperature, air speed and steam pressure are used. The PMV can be obtained using Eq. 2,

$$PMV = (0.303e^{-0.036M} + 0.028) \cdot \{(M - W) - 3.05E^{-3}[5733 - 6.99(M - W) - P_w] - 0.42[(M - W) - 5815] - 1.7E^{-5}(5867 - P_a) - 0.0014M(34 - T_a) - 3.96E^{-8} \cdot f_{cl}[(T_{cl} + 273)^4 - (T_r + 273)^4] - f_{cl}h_c(T_{cl} - T_a)\} \tag{2}$$

where M is the metabolic rate (W/m^2), W is the external work (W/m^2) of the body area, P_w is the partial water pressure (Pa), T_a is the air temperature (ºC), f_{cl} is the ratio of clothed surface to nude surface, T_{cl} is the clothing temperature (ºC), T_r is the mean radiant temperature (ºC) and h_c the convective heat transfer coefficient (W/m^{2}K), [10, 11].

PPD, which represents the percentage of people living in uncomfortable conditions, can be estimated by Eq. 3.

$$PPD = 100\text{-}95e^{-(0.03353PMV^4+0.2179PMV^2)} \quad (3)$$

Indoor Air Quality

The amount of CO_2 in the indoor air was used to determine the indoor air quality of the shelter. The concentration indicates the ventilation rate needed within a room. A certain time is needed to reach the equilibrium level, and this time depends on the ventilation rate, the number of people occupying the shelter and their activity. At the building location, the outdoors levels of CO_2 are relatively constant and range between 350 and 600 ppm, which implies that, in a naturally ventilated building, the indoor levels will never be below the outdoor level, [6-10].

Building External Layout

At the location, the 204 homes should be placed on 96 lots. The building strategy was not defined, and the homes were placed in a regular basis. Due to the lack of HVAC systems, the aim of this paper was to optimize both the distribution of homes and the building materials to obtain higher thermal comfort with a very low cost and provide an alternative option to traditional shelters.

To optimize natural ventilation in the homes, two different distributions were studied using the external CFD module in the Design Builder software [12]. The external CFD study provides useful information about the air pressure, velocity and distribution outside the buildings. The first configuration studied was a regular one in which the homes were placed at regular intervals. This layout could be the most easily designed but not optimal for achieving the highest possible comfort levels. To evaluate the influence on the natural ventilation, the homes were distributed in non-regular intervals, and the model was studied again with the Design Builder software [12].

Building Internal Layout

The purpose of the homes is to provide a secure shelter for the IDPs and to be the basis for the future development of a new city in the location. Given this strategy and the lack of monetary resources, the buildings have only an internal division and no separated zones or areas. The lack of HVAC systems, electricity, water or primary energy implies that homes will have no facilities. The houses will be used as living and sleeping rooms for their inhabitants. The useful surface of each house is up to 28 m^2, which is distributed as shown in Figure 4 and Figure 5.

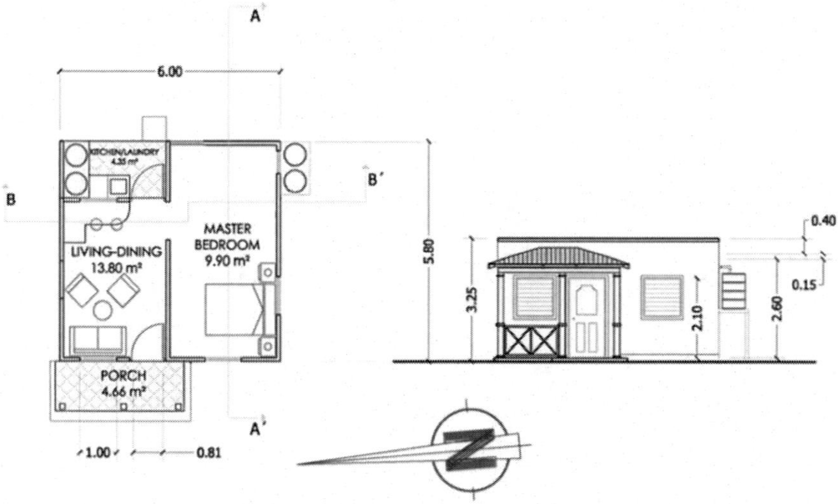

Figure 4. Building plant and sections.

Figure 5. Building sections.

When the economic conditions improve in the location, the homes could be used as the first floor of a larger home intended for a family. This solution provides a modular and time-scalable solution that could serve as the basis for a city, as shown in the model in Figure 6.

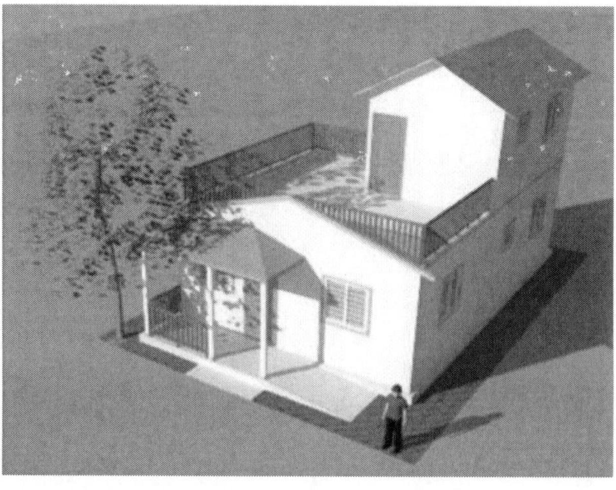

Figure 6. Building modular construction.

Building Occupation

Due to the lack of shelters, the building should be optimized for a high occupancy level. The occupation will be 10 inhabitants, and the occupancy will be distributed as shown in Table 3.

Table 3. Occupancy during the day

Time	Inhabitants	Density (people/m^2)
7:00–12:00	10	0.36
12:00–19:00	5	0.18
19:00–7:00	10	0.36

It is necessary to improve ventilation to increase both comfort levels and indoor air quality because of the high occupancy levels during the day. The ventilation requirements have been solved through natural ventilation. Natural ventilation provides a passive cooling technique to improve both thermal and quality air conditions.

Building Materials

Due to the lack of monetary resources in the zone, low cost construction techniques are required. It is necessary to use building materials that are easily placed and do not require skilled manpower because of the absence of electricity and energy at the location of the lots.

In this case, the solution is provided by using concrete as the building material. The concrete will be provided by Republica Dominicana, and the buildings will be constructed within a short period. Concrete provides the required low cost and high durability. It also allows future enlargement of the building surface by adding an additional floor. The concrete typology is different for the walls and for the roof. The concrete used in the walls is conventional concrete due to the need for a high resistive capacity. In the roof, lightweight concrete is chosen to lower the roof weight, but it provides sufficient structural resistance to allow future building enlargement.

The physical characteristics of the materials used are shown in Table 4. The windows are formed of 4 mm simple glazing with a U value of $U_{glaz} = 6$ (W/m^{2}K) and an aluminum frame, where $U_{frame} =$ 5.7 (W/m^{2}K). The materials used provide a low cost solution and can be bought in the local market from nearby countries. The doors are aluminum, and $U_{doo} = 2$ (W/m^{2}K). The sizing of the windows is based on two opposed objectives: to ensure optimal natural ventilation and to reduce heat gains into the building. To achieve the best possible solution, the windows are placed oppositely in the building to ensure cross ventilation. The height for the location of the windows is selected according to natural lighting criteria. Cross ventilation relies on wind to force cool exterior air into the building through the windows and to force warm interior air out of the building through an outlet, which, in this case, is another window. Using double or triple glazing reduces heat gains, but its use is not possible due to its high cost and low availability at the location. The theoretical section of the building is 1 m^2, but a metal grid is emplaced to both reduce direct solar gains and prevent insects from entering the building. The operable section is used as an input in the EnergyPlus model and is set to 60% of the section, which implies an operable section of 0.6 m^2 for each window. The materials described above are selected according to the money available for the building construction. As mentioned before, the aim of this paper is to provide a low-cost solution to improve the thermal comfort in the buildings. To achieve the required comfort, passive cooling technologies are used and studied.

Table 4. Concrete physical characteristics

Light weight concrete	
Conductivity [W/(mK)]	1.8
Specific weight [kg/m^3]	1,600
Specific heat [kJ/(kgK)]	0.75
Emissivity	0.9
Reinforced concrete	
Conductivity [W/(mK)]	1.8
Specific weight [kg/m^3]	2,400
Specific heat [kJ/(kgK)]	0.278
Emissivity	0.93

PASSIVE COOLING TECHNOLOGIES USED

To improve the thermal behavior of the shelters, seven study cases are proposed, including some variations of the base case in which the building is of simple concrete and no passive cooling is used. The solution provided should have a cost below \$100/m^2. Due to the low budget, four techniques are studied:

- Expanded Polystyrene Insulation (EPS) in the roof
- Natural ventilation through windows
- Improved natural ventilation with a roof opening
- Cool roof

Using the conductivity values and the thermal resistance, the U values are calculated for the base case configuration for the wall and the roof and shown in Table 5.

The combination of some of these technologies provides a cheap solution and increases the thermal comfort for inhabitants. Cool roofs have been referenced in many studies, but their potential in low cost buildings for IDPs has not been evaluated yet.

Table 5. Thermal values

Roof materials				
Description	Layer	Thickness	Conductivity	Thermal conductance
	Number	mm	W/mK	$(W/m^2)/K$
EPS Roof Case (EPS roof)				
Exterior film coefficient				56.700
Light weight concrete ($1600\ kg/m^3$)	1	60.0	0.750	12.500
EPS	2	84.0	0.029	0.341
Light weight concrete ($1600\ kg/m^3$)	3	60.0	0.750	12.500
Interior film coefficient				9.300
U-value-nominal				0.311
Base Case Roof (BC Roof)				
Exterior film coefficient				56.700
Light weight concrete ($1600\ kg/m^3$)	1	60.0	0.750	12.500
EPS	2	0.0	0.029	0.000
Light weight concrete ($1600\ kg/m^3$)	3	60.0	0.750	12.500
Interior film coefficient				9.300
U-value-nominal				3.508
Walls Concrete (WC)				
Exterior film coefficient				33.40
Light weight concrete ($2400\ kg/m^3$)	1	100	2.300	0.043
Interior film coefficient				8.300
U-value-nominal				0.193

Cool Roof

A cool roof is a building typology that reduces the cooling demand. Cool roofs have surfaces intended to reflect sunlight and emit heat more efficiently than hot or dark roofs, which allows them to remain cooler in the sun. Solar reflectance and thermal emittance are the two key material surface properties that determine the temperature of a roof; in this case, the material used is concrete. The emittance and reflectance range from 0 to 1, and the larger these two values are, the cooler the roof will remain in the sun. Reducing the temperature of the roof with a cool roof can also increase the need for heating during cold seasons; however, in this case, the location has no heating needs, and thus the solution is optimal [13-20].

The materials used in the cool roof are a particular category of materials and components that are able to stay cool under solar radiation. They are characterized by high solar reflectance values, which reduce the solar radiation absorbed by conventional building materials and limit the surface temperature increase in the presence of high solar loads. In this case, the specific roof surface of the shelter is very high compared with the useful plant surface, with 1 m^2 of roof surface per m^2 of useful building plant. This high value of the ratio between the roof and plant surface means that reducing thermal gains in the roof is a priority, [21-22].

In the shelter, the cool roof characteristics allow the material to increase its surface temperature only a few degrees above the ambient temperature [23-29], reducing the heat released from the roof to the indoor air by convection and reducing the indoor temperature [30-35]. Lower surface temperatures also reduce the heat transfer from the roof to the built environment, which reduces the cooling load of the building. Most widely used cool roof materials are available as tiles, asphalt shingles, metal roofs, elastomeric coatings, and single-ply membranes. The most common cool roof system is a white coating. The results of the review and testing of such materials can be found in many studies. The studies show that metal roofs, even if characterized by low emittance values, can also be used as cool roofs in the case of high solar reflectance [36]. Cool colored materials have recently been developed and tested. They are mainly based on spectrally selective inorganic or organic pigments that are characterized by a high reflectance in the infrared portion of the solar spectrum but that maintain the reflectance profile of conventional colored materials in the visible spectrum [37]. They are used for coatings and tiles and present a solar reflectance that is much higher than that of conventional colored materials [38-40].

Table 6. Cool roof paint

Reflectivity	0.9235
Thermal conductivity acrilyc paint (W/(mK))	0.5
Thermal conductivity acrilyc paint with additive (W/(mK))	0.1292

The solution selected in the case studies is a nanotechnological and microtechnological additive that makes any acrylic, water-based paint thermally insulating and water repellent. The additive makes it possible to improve the thermal behavior with low cost, and it is easy to use at the building location. When the additive is added to paint, the reflectivity value becomes 0.9235, and it reduces the thermal conductivity of typical acrylic paints by a factor of four, to a value of 0.1292 W/mK according to EN 12667, as shown in Table 6.

Improved Natural Ventilation and Internal CFD Optimization

Natural ventilation was simulated to improve the thermal behavior by means of a roof opening intended for heat evacuation. This option was compared with natural ventilation using windows and with the combined use of both systems. Natural ventilation in the building was optimized and simulated using the EnergyPlus CFD calculation capabilities integrated in Design Builder. For each case study, the roof design, roof opening size and roof opening location were optimized using CFD simulations.

Case Studies

The combination of the technologies provides eight study cases that were simulated and evaluated, as listed in Table 7. The materials are shown in Table 3. The base case building typology for the walls is based on walls of concrete (WC), and the roof has two different configurations, a base concrete roof (BC Roof) and a base concrete roof with EPS insulation (EPS Roof). For both roof configurations, the cool roof paint additive is evaluated (CR).

Table 7. Case studies

Case	Name	Description	Walls	Roof
1	BASE CASE	Base case, windows natural ventilation as passive technique	WC	BC Roof
2	BASE CASE OPENING	Base case, windows natural ventilation and upper opening as passive technique	WC	BC Roof + roof opening
3	BASE CASE COOL ROOF	Base case, windows natural ventilation as passive technique and cool roof	WC	BC Roof + CR
4	BASE CASE COOL ROOF OPENING	Base case, windows natural ventilation, upper opening as passive technique and cool roof	WC	BC Roof + CR + roof opening
5	EPS CASE	EPS roof insulation, windows natural ventilation as passive technique	WC	EPS Roof
6	EPS CASE OPENING	EPS roof insulation, windows natural ventilation and upper opening as passive technique	WC	EPS Roof + roof opening
7	EPS CASE COOL ROOF	EPS roof insulation, windows natural ventilation as passive technique and cool roof	WC	EPS Roof + CR
8	EPS CASE COOL ROOF OPENING	EPS roof insulation, windows natural ventilation, upper opening as passive technique and cool roof	WC	EPS Roof + CR + roof opening

Simulation

The shelters were modeled for each of the eight case studies and simulated using EnergyPlus. The OpenStudio software package [41] was used as a graphical interface through Google SketchUp [42]. The weather conditions were converted from a standard TMY2 model to the EnergyPlus weather file format. Figure 7 and Figure 8 show the temperature conditions and solar radiation at the location, respectively.

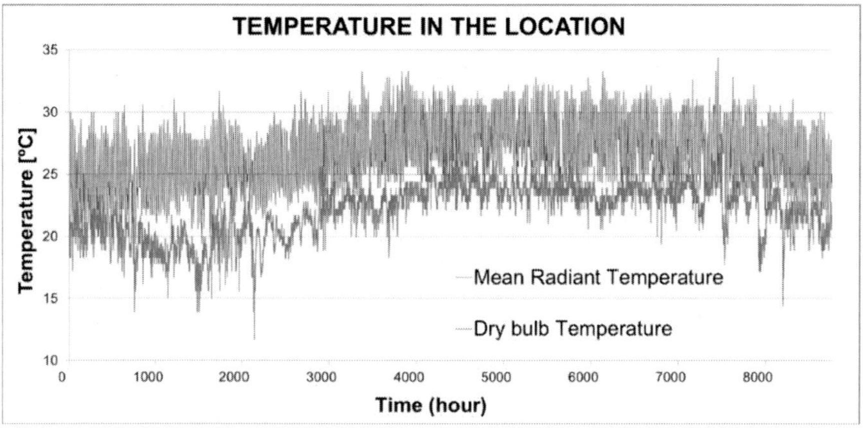

Figure 7. Hourly temperatures.

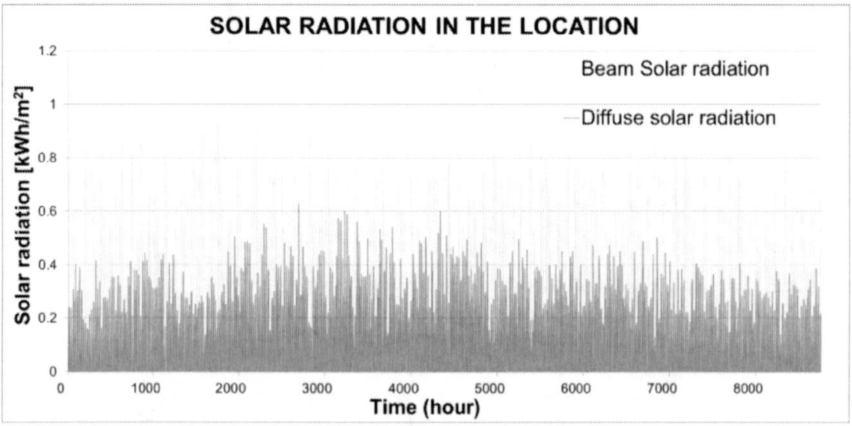

Figure 8. Hourly solar radiation

The values are shown for a typical meteorological year and are discretized using hourly values. For each case study, comfort values were used to determine the effectiveness of natural ventilation using window openings and a roof opening. Different roof opening locations were simulated, and the PMV was calculated for the design day for each location. The PMV values, temperature distribution, air velocity and comfort conditions (T_{op}, T_r) were simulated for this design day.

RESULTS AND DISCUSSION

An hourly simulation was performed for each of the eight case studies, and the thermal and comfort parameters were studied for each of them. The financial parameters are necessary to inform decisions about possible investment in the required building materials, and thus the cost of each of the solutions is calculated and compared. An economic and financial study was conducted. For the paint additive for the cool roof, the cost was studied based on data from a local provider in Republica Dominicana, the nearest country from which the building materials could be delivered. Additionally, the United States Department of Energy Cool Roof guide was consulted. The paint additive reduces the initial investment cost up to 50% in comparison with other traditional systems such as coatings [43]. The results were studied on a monthly basis, and the typical cooling day was used to compare the results for each of the cases, as shown in Table 8. The design day allows for analysis and a decision on the optimal solution to be applied.

Table 8. Case studies main results

Case study	Monthly results			Cooling design day				Economic analysis						
	Air temperature	Mean radiant temperature	Operative temperature	Relative humidity	Air temperature	Mean radiant temperature	Operative temperature	Building typology	Unitary cost increase	Unit	Ventilation system	Ventilation system cost increase	Total cost increase	Specif cost increase
	°C	°C	°C	%	°C	°C	°C	Type	D$	m²	Type	$	$	$/m²
CASE 1	27.08	29.12	28.10	69.98	28.79	30.05	29.42	BC	0.00	28.00	Windows	$0	$0	$0
CASE 2	27.04	29.11	28.08	69.98	28.82	30.08	29.45	BC	0.00	28.00	Windows + roof opening	$123	$123	$4
CASE 3	26.71	27.11	26.91	77.99	26.35	26.42	26.38	BC + CR	37.00	28.00	Windows	$0	$1036	$37
CASE 4	26.71	27.15	26.93	79.54	25.99	25.84	25.91	BC + CR	37.00	28.00	Windows + roof opening	$123	$1159	$41
CASE 5	26.71	27.00	26.85	78.16	26.30	26.33	26.32	EPS	41.00	28.00	Windows	$0	$1148	$41
CASE 6	26.71	27.05	26.88	77.84	26.39	26.44	26.41	EPS	41.00	28.00	Windows + roof opening	$123	$1271	$45
CASE 7	26.69	26.90	26.80	78.74	26.14	26.06	26.10	EPS + CR	78.00	28.00	Windows	$0	$2184	$78
CASE 8	26.70	26.96	26.83	78.39	26.23	26.18	26.21	EPS + CR	78.00	28.00	Windows + roof opening	$123	$2307	$82

Table 9. Case results

	Relative humidity	Air temperature	Mean radiant temperature	Operative temperature
	%	°C	°C	°C
Max value	79.54	28.82	30.08	29.45
Case for max value	CASE4	CASE2	CASE2	CASE2
Min value	69.98	25.99	25.84	25.91
Case for min value	CASE1	CASE4	CASE4	CASE4
Δ	9.55	2.83	4.25	3.54
% Δ MAX–MIN	14%	11%	16%	14%

For each case, a detailed analysis was conducted, and design day simulation results delivered from EnergyPlus were studied using a spreadsheet. The detailed heat gains results for the building surfaces allow for the study of the efficiency of each demand reduction system, especially the cool roof. Case 4, using cool roof technology and a roof opening, has the best performance, as shown in Table 9. The cost for the proposed technology was calculated in $\$/m^2$, enabling analysis of the cost increase for each proposal. For the cool roof, Case 4, the cost increase is $\$37/m^2$, which is an acceptable value for the project. The total cost increase in this case is $1,159.

The best results are achieved with the combined use of a cool roof and a natural ventilation strategy. All of the results show that the EPS system used as a stand-alone solution does not provide a good solution. According to the design day and parameters and the whole year simulation results, its use lowers comfort values. The low thermal comfort occurs because the air temperature increase in the building space combined with the low conductivity value of the roof leads to indoor heat accumulation. Natural ventilation increases the relative humidity, but high occupancy levels in the building require higher values for air change. The design day temperature values are shown in Figure 9, and the design day relative humidity is shown in Figure 10.

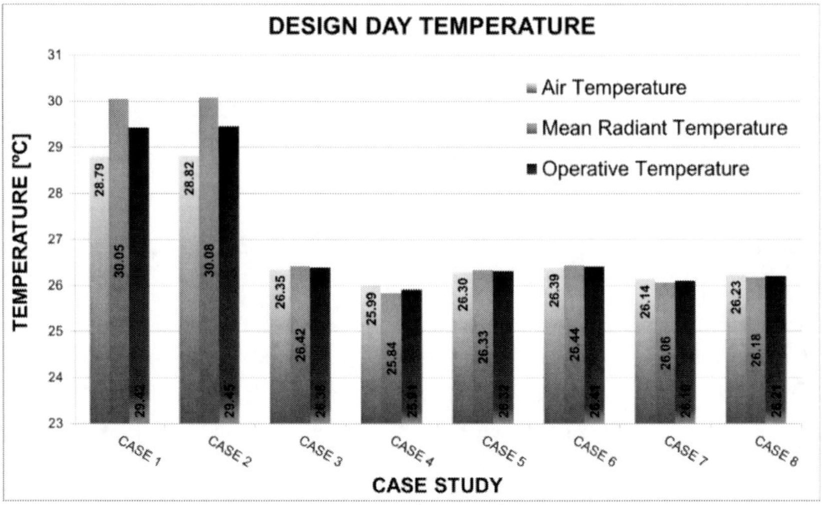

Figure 9. Design day temperature.

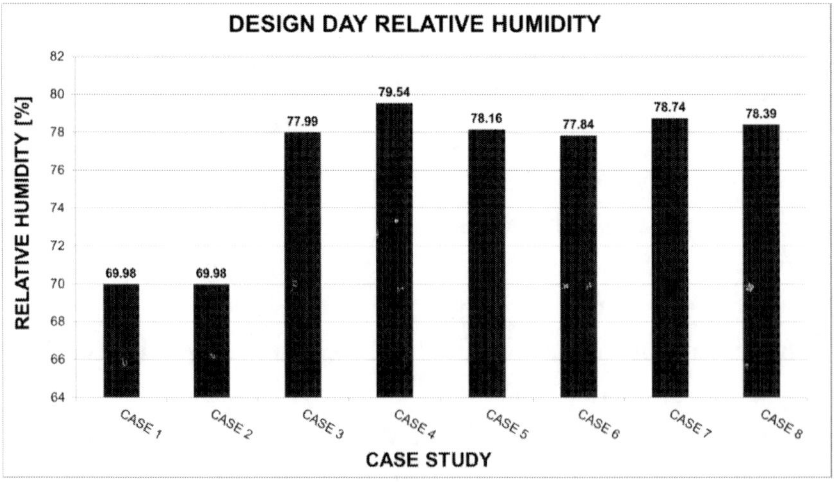

Figure 10. Design day relative humidity.

The main objective of the proposed option is to increase thermal comfort. The thermal comfort is evaluated in the simulation by means of T_{op} and T_r. Both temperatures are analyzed because they are important components of thermal discomfort. The solution provides a significant increase in comfort (T_{op}) and reduces T_r, leading to thermal conditions in the shelter that are much more comfortable. The T_{op} value over the course of a day is shown in Figure 11. The cool roof reduces heat accumulation in the building structure, which lowers temperatures during the last hours of the day. A conventional thermal analysis using a dry bulb temperature would not consider all of these relevant aspects of thermal comfort; thus, dynamic thermal analysis is an extremely useful tool in this type of analysis or for applications that are specifically intended for people in very bad living conditions. The results show that a design using T_a as the design criterion would lead to a very poor level of indoor comfort. A good value in terms of air temperature can be achieved, but thermal comfort is not reached, especially in the case of high occupancy levels. The minimum indoor air quality was set according to occupancy levels as a requirement for natural ventilation. The simulation shows that adequate values of indoor air quality are achieved by means of natural ventilation.

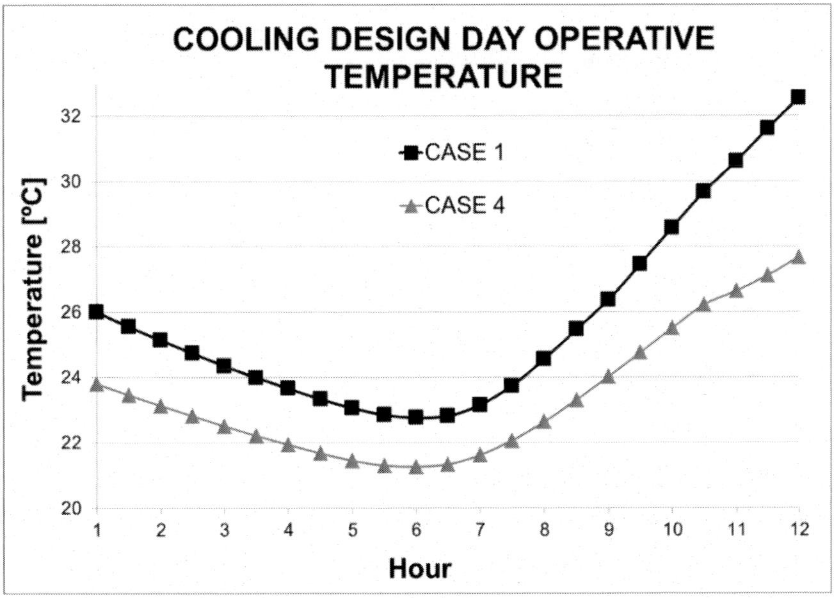

Figure 11. Cooling design day operative temperature.

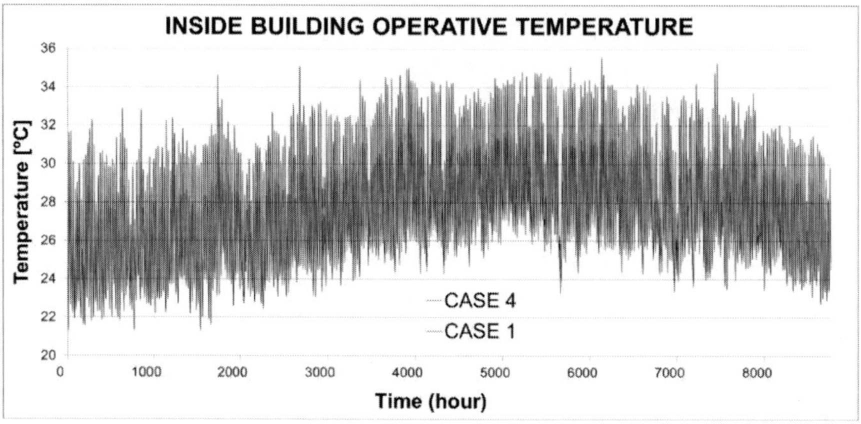

Figure 12. Operative temperature along the year.

To extend the results for the entire year, the hourly results were analyzed. The indoor T_{op} is shown in Figure 12. Case 4 provides better results for the whole year than case 1. The results for a whole year show that, for the climatic conditions of the placement, the thermal parameters are improved for the whole year rather than only for the design day. The simulation results suggest that case 4 is the most suitable option for the required comfort needs of the building, but a prototype should be built to validate the results. Due to the low cost of the proposed solutions and the high need of new building construction in Haiti, using passive techniques should be integrated as part of the design criteria. This strategy would increase the thermal comfort for the inhabitants and provide useful information for research purposes and continuous improvement.

To analyze the behavior during different months and detect variability over the year, the hourly values were analyzed on a monthly basis, showing that the combined solution of a cool roof and improved ventilation techniques provides the best indoor conditions over the whole year. The air temperature is lower, but T_r and T_{op} are also lower, which leads to much more comfortable indoor conditions, as shown in Figure 13. The analysis of T_a, T_r and T_{op} shows that a very low cost solution and an exhaustive design strategy can ensure better conditions for people living inside the building.

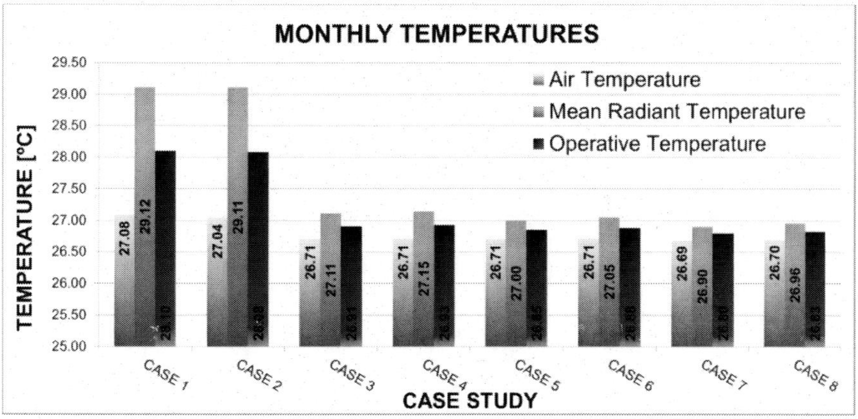

Figure 13. Design day relative humidity.

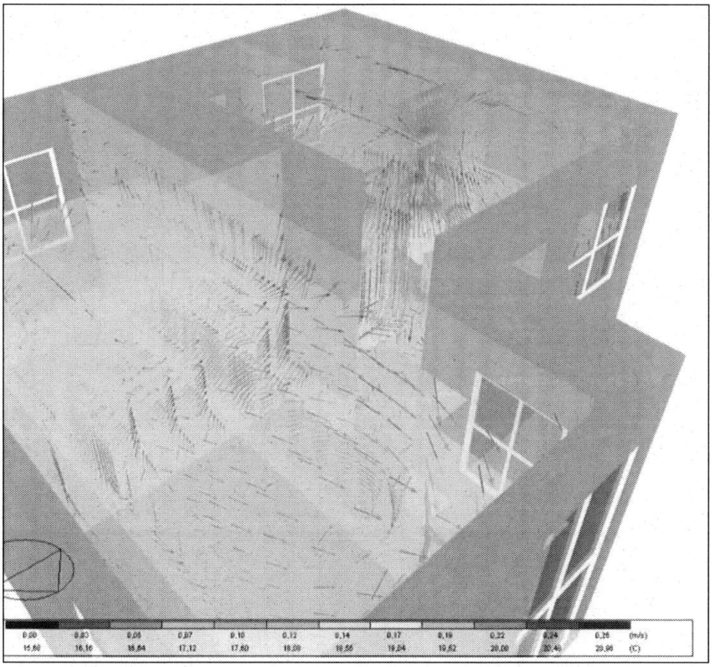

Figure 14. Design day internal CFD temperature and air velocities.

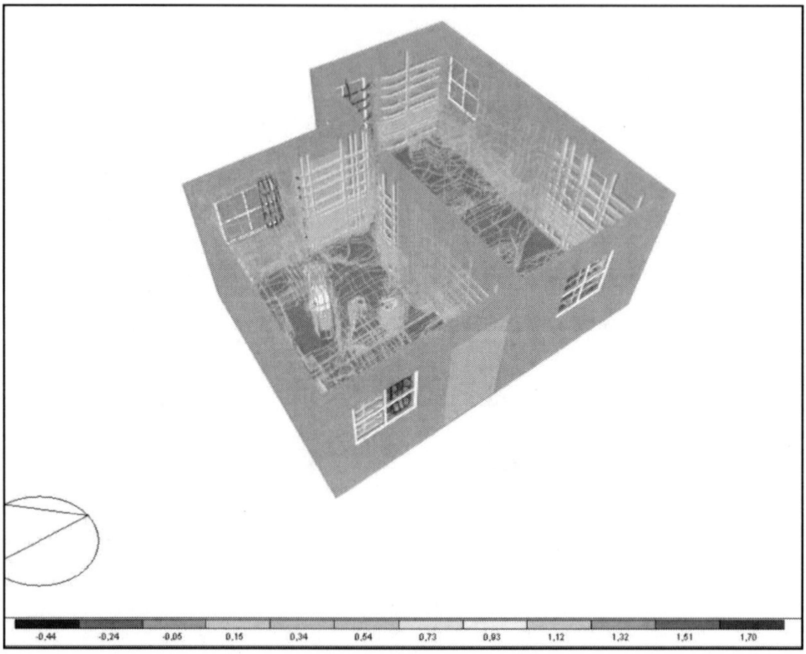

Figure 15. Design day internal CFD PMV values.

To validate the solution, other comfort value results for the building and CFD analysis results for the design day were analyzed. The PMV values and PPD values are monitored in the living room of the building to find the solution that leads to better parameters. For each proposed case study, the CDF, PMV and PPD values were evaluated to analyze the thermal behavior. The combination of natural ventilation, a cool roof and a roof opening increases comfort values, which corresponds to the previously stated results that show a reduction in T_r and T_{op} for case 4. Figure 14 shows a detail of the internal CFD analysis results for the design day for case 4. The air temperature and air velocity are shown on a color scale. The roof opening, where air velocities increase, is key to increase air circulation inside the building and heat evacuation. Figure 15 shows a detail of the PMV values from the same analysis. The results of the CFD comfort analysis were evaluated and graphically analyzed to ensure that case 4 is the best available solution. The analysis, as shown in the figure, is quite useful to detect the causes of low thermal comfort. High T_r values are an important cause of discomfort in hot and humid climates. The figure shows that higher values are obtained for the placements near the walls and openings while zones with adequate air circulation have better comfort parameters, which suggests that CFD analysis is a powerful and low cost tool for low cost building improvement.

CONCLUSION

Globally, many people live as IDPs. The typical buildings in IDP camps are tents. In this paper, a low-cost building using low-cost techniques to increase thermal comfort is proposed and validated. A concrete building combined with passive cooling techniques can increase thermal comfort for the inhabitants with no HVAC or energy facilities. The proposal is intended for climates with high temperatures where cooling systems are needed to achieve thermal comfort. For validation, the solution is simulated and studied for Haiti. In Haiti, there are now more than 1.3 million people living in tents or in transitional shelters. Low cost is a must in the country, but engineering techniques can be used to improve thermal comfort for the people living inside buildings. At the location, no facilities exist; thus, no HVAC or energy is provided. Consequently, this research fully matches the requirements. The lack of facilities makes passive cooling systems the only way to improve thermal conditions in the shelters.

The results of the simulations shown in the paper lead to the optimal solution for cooperation development. Required low cost and lack of building materials make concrete buildings the best solution because their constituent materials can be easily bought in adjacent countries, and no difficult construction equipment is required. The occupancy levels are very high, which makes thermal conditions very poor. A solution for this problem is provided and validated for the hot and humid conditions of Haiti. The simulation results show that low-cost cool roofs provide an effective solution that, in combination with natural ventilation, increases thermal comfort up to 16%, but a prototype should be built to validate the results. Higher values can be achieved at locations with better climatic conditions and lower occupancy levels. The system will allow future high comfort conditions for families occupying the building after transitional shelter use. For comparison and validation, 6 other case studies are evaluated in the research, including EPS insulation and the combination of all considered technologies. The analysis showed that the best results are achieved with cool roofs. Cool roofs are usually referenced as a way to reduce energy consumption in cities, but they can provide a solution to improve living conditions for many millions people living in shelters. The cost increase is less than $50/m², which makes their use affordable and extremely low cost in comparison to other building systems.

REFERENCES

[1] Myers EG. Effect of ventilation rate and board loading on formaldehyde concentration: a critical review of literature. *Forest Products J* 2004; 34:10.

[2] de Dear RJ. A global database of thermal comfort field experiments. *ASHRAE Trans* 1998;104(1):1141–52.

[3] Auliciems A. Towards a psycho-physiological model of thermal perception. *Int J Biometeorol* 1981;25:109–22.

[4] Humphreys MA. Field studies of thermal comfort compared and applied. *J Inst Heat Vent Eng* 1975;44:5–27.

[5] Baker NV, Standeven MA. A behavioural approach to thermal comfort assessment in naturally ventilated buildings. In:

Proceedings CIBSE national conference. Eastbourne, Chartered Institute of Building Services Engineers, London; 1996. p 76–84.

[6] Humphreys MA. The influence of season and ambient temperature on human clothing behaviour. In: Fanger PO, Valbjorn O, editors. *Indoor climate*. Copenhagen: Danish Building Research; 1979.

[7] de Dear RJ, Brager GS. Thermal comfort in naturally ventilated buildings: revisions to ASHRAE Standard 55. *Energy and Buildings* 2002;34(6):549–61.

[8] Humphreys MA, Nicol JF. Understanding the adaptive approach to thermal comfort. *ASHRAE Trans* 1998;104(1):991–1004.

[9] Olesene BW. A better way to predict comfort. *ASHRAE J,* August 2004, p. 20–26.

[10] Markov D. *Practical evaluation of the thermal comfort parameters*. Bulgaria: Technical University of Sofia; 2006.

[11] Roulet CA. Simple and cheap air change rate measurement using CO2 concentration decays. *Air Infiltration and Ventilation Centre technical note*; 1991. p. 34.

[12] *Design Builder*, http://www.designbuilder.co.uk/ accessed 21/02/2012.

[13] Bertoldi P, Atanasiu B. Electricity consumption and efficiency trends in European Union, European Commission. Joint Research Centre. *Institute for Energy Status Rep*; 2009.

[14] Wilkes KE. *Model for roof thermal performance*. ORNL; 1989 [ORNL/CON- 274].

[15] Bretz S, Akbari H. *Durability of high-albedo roof coatings and implications for cooling energy savings*. Lawrence Berkeley Lab; 1994 [LBLe34974].

[16] Winandy JE, Beaumont R. *Roof temperatures in simulated attics*. Forest Products Laboratory; 1995 [Res. Pap. FPL-RP-543].

[17] Parker DS, Barkaszi Jr SF, Chandra S, Beal DJ. *Measured cooling energy savings from reflective roofing systems in Florida: field and laboratory research results*. Florida Solar Energy Center

(FSEC) and Lawrence Berkeley National Laboratory; 1995 [FSEC-PF-331-98].

[18] Parker DS, Sherwin JR, Sonne JK, Barkaszi Jr SF. *Center, demonstration of cooling savings of light colored roof surfacing in Florida commercial buildings: our Savior's School*. Florida Solar Energy Center; 1996 [Rep. No. FSEC-CR- 904-96].

[19] Konopacki S, Akbari H, Gartland L. *Cooling energy savings potential of lightcolored roofs for residential and commercial buildings in 11 US metropolitan areas*. Lawrence Berkeley Lab; 1997 [LBNL-39433-Exec.Summ].

[20] Parker DS, Huang YJ, Konopacki S, Gartland LM, Sherwin JR, Gu L. Measured and simulated performance of reflective roofing systems in residential buildings. *Transactions-American Society of Heating Refrigerating and Air Conditioning Engineers* 1998;104:963-75.

[21] Bretz S, Akbari H. Long-term performance of high-albedo roof coatings. *Energy and Buildings* 1997;25:159-67.

[22] Rosenfeld A, Akbari H, Romm J, Pomerantz M. Cool communities: strategies for heat island mitigation and smog reduction. *Energy and Buildings* 1998;28:51-62.

[23] Akbari H, Konopacki S, Pomerantz M. Cooling energy savings potential of reflective roofs for residential and commercial buildings in the United States. *Energy* 1999;24:391-407.

[24] Petrie TW, Atchley JA, Childs PW, Desjarlais AO. Effect of solar radiation control on energy costs e a radiation control fact sheet for low-slope roofs, *Proceedings of the performance of the exterior envelopes of whole buildings VIII: Integration of building Envelopes*. Paper 146. American Society of Heating, Refrigerating and Air Conditioning Engineers, Inc; 2001.

[25] Akbari H, Levinson R, Miller WA, Berdahl P. Cool colored roofs to save energy and improve air quality, *International Conference "Passive and low energy cooling for the built environment"*; 2005. Santorini, Greece89-100.

[26] Shashua-Bar L, Tsiros I, Hoffman M. A modeling study for evaluating passive cooling scenarios in urban streets with trees e case study: Athens, Greece. *Building and Environment* 2010. doi:10.1016/j.buildenv.2010.06.008.

[27] Levinson R, Akbari H, Konopacki S, Bretz S. Inclusion of cool roofs in nonresidential Title 24 prescriptive requirements. *Energy Policy* 2005;33:151-70.

[28] Parizotto S, Lamberts R. Investigation of green roof thermal performance in temperate climate: A case study of an experimental building in Florianópolis city, Southern Brazil. *Energy and Buildings* 2011;43(7):1712–1722.

[29] Levinson R, Berdahl P, Asefawbrerhe A, Akbari H. Effects of soiling and cleaning on the reflectance and solar heat gain of a light-colored roofing membrane. *Atmospheric Environment* 2005;39:7807-24.

[30] Wong NH, Chen Y, Ong CL, Sia A. Investigation of thermal benefits of rooftop garden in the tropical environment. *Building and Environment* 2003;38(2):261–70.

[31] Xu T, Sathaye J, Akbari H, Garg V, Tetali S. Quantifying the direct benefits of cool roofs in an urban setting: reduced cooling energy use and lowered greenhouse gas emissions. *Building and Environment* 2012;48:1-6.

[32] Synnefa A, Santamouris M, Akbari H. Estimating the effect of using cool coatings on energy loads and thermal comfort in residential buildings in various climatic conditions. *Energy and Buildings* 2007;39:1167-74.

[33] Akbari H, Menon S, Rosenfeld A. Global cooling: increasing world-wide urban albedos to offset CO_2. *Climatic Change* 2008;94:275-86.

[34] Kültür S., Türkery N. Assessment of long term solar reflectance performance of roof coverings measured in laboratory and field. *Building and Environment* 2012;48:164-172.

[35] Halwatura R, Jayasinghe M. Influence of insulated roof slabs on air conditioned spaces in tropical climatic conditions—a life cycle cost approach. *Energy and Buildings* 2009;41:678–686.

[36] Levinson R, Akbari H. Potential benefits of cool roofs on commercial buildings: conserving energy, saving money, and reducing emission of greenhouse gases and air pollutants. *Energy Efficiency* 2009;3:53-109.

[37] Menon S, Akbari H, Mahanama S, Sednev I, Levinson R. Radiative forcing and temperature response to changes in urban albedos and associated CO_2 offsets. *Environmental Research* Letters 2010;5:014005.

[38] Urban BJ, Roth K. Guidelines for selecting cool roofs. U.S. Department of energy. Building technologies program and Oak Ridge National Laboratory. *Building Technologies Program;* 2010. 1-23.

[39] Boixo S, Diez-Vicente M, Colmenar A, Alonso M. Potential energy savings from cool roofs in Spain and Andalusia. *Energy* 2012; 38(1);425-438.

[40] C. Romeo, M. Zinzi. Impact of a cool roof application on the energy and comfort performance in an existing non-residential building. *A Sicilian case study. Energy and Buildings*, In Press, Corrected Proof, Available online 28 July 2011.

[41] *OpenStudio Suite*, http://openstudio.nrel.gov/ accessed 02/ 02/2012.

[42] *Google SketchUp* http://sketchup.google.com/ accessed 12/ 02/2012.

[43] *Guidelines for Selecting Cool Roofs*. U.S. Department of Energy, Building Technologies Program, 2010.

In: Alternative Energy Systems … ISBN: 978-1-53614-203-7
Editors: A. Colmenar-Santos et al. © 2019 Nova Science Publishers, Inc.

Chapter 4

SOLAR THERMAL SYSTEMS FOR HIGH CONSUMPTION DEMAND BUILDINGS

***Julio Vale-Vale*[1], *Carlos de Palacio*[2] *and Pedro-Miguel Ortega-Cabezas*[3,*]**

[1]Departamento de Ingeniería Eléctrica, Electrónica, Control, Telemática y Química Aplicada a la Ingeniería, Universidad Nacional de Educación a Distancia (UNED), Madrid, Spain

[2]ABB, Madrid, Spain

[3]PSA, Madrid, Spain

ABSTRACT

Solar thermal has been consolidated as a renewable energy solution for new buildings during recent years. Although this system has been easily integrated in low-rise buildings, a continuous challenge is the integration of solar thermal into the demanding architectural solutions for high-rise buildings. This article presents solutions for solar water

[*] Corresponding Author Email: pm.ortega-cabezas@gmail.com.

installations in high rise buildings. It describes the integration of solar collectors into the building, hot water distribution installation and proposes a solution to minimize the risk of exposure to Legionella. As an example of the requirements of solar thermal in developing countries, the article analyses the development of the solar hot water standards in Brazil, where although there is a draft document for a federal standard, cities like Sao Paulo have already in place solar thermal requirements for any new building in the city. In order to promote the development of renewable integration, initiatives like PROCEL certify buildings energy performance if they meet the improved target, compared to regulatory requirements, of generating a minimum of 60% of hot water using solar thermal. The presented case study for a 5-star hotel clearly justifies the installation of solar thermal systems in buildings with a high demand for hot water.

Keywords: integrated solar thermal, high rise buildings, solar hot water, Legionella, PROCEL

INTRODUCTION

The solar energy used to generate heat is one of the oldest solar energy resources. However, it was necessary to wait until the mid-1950s to see the first commercial building with a solar thermal solution. The architect Frank Bridgets designed the first world commercial building using solar water heating and passive design [1].

In Brazil, the first solar thermal systems were installed in the 1970s, but it was not until the 1990s when the systems start to be installed in a professional way, in answer to a demanding market and the development of the Brazilian legislations and standards ABNT for the solar thermal sector.

Although solar thermal has been used the last years in dwellings, at present, due to the new regulations, the applications of solar water heater in new construction buildings is compulsory. This is giving a boost to the solar thermal market.

There are two major issues related to the integration of solar water heating system in buildings. One is the integration of the solar heating collectors into the architectural design. The amount of solar collectors required to cover the water demand required in the regulations require a large space, so the roof of the buildings are not providing enough space for the number of collectors required, therefore the facades of the buildings and surrounding areas need to be explored for the location of the collectors. This issue, that is becoming really difficult for the architects, is even more important in buildings with large water consumption like hotels or hospitals.

The second issue is the integration of the technical equipment (water tank, pumps, pipelines, etc.) within the building. This is not a major problem in low rise building where the water heater plant is associated to the gas or electrical support water heater and the pipes are shared, however it is more challenging for high rise buildings where the height of the building affects the pressure in the vertical pipes, requiring specific solutions.

Due to all these challenges, this paper will focus on the study of solar thermal systems for high rise buildings based on a case study of a proposed hotel in Brazil, in the highly populated city of Sao Paulo

HIGH RISE BUILDINGS

With the development of the urban residential construction, growing due to the growth of the economy in Brazil and people moving into the cities, the urban population has been rising in the cities in Brazil, encouraging a fast development in the building industry [2].

The beginning of the high rise in Brazil was in 1920s, promoted by an import-substituting industrialization, with special impact in Sao Paulo, but it was in the last two decades when it became important either in cities where high rise building existed before and cities where they were not [3].

For cities like Sao Paulo with a city center where the price of the land is growing continuously, the high rise construction is the way to build profitable developments, building upwards as to minimize the cost of the land as per the total area of the building. In this way, the same land can provide more offices, apartments, commercial areas, hospitals or hotel rooms.

This is the reason why in the 1980s and 1990s a good number of high rise office buildings were built in Sao Paulo. The growing economy in Brazil and Sao Paulo working as a business hub for Latin America, invited business people to set up their offices in this city. As a consequence high rise building hotels were built to provide accommodation to the international visitors.

Hotel high rise buildings have large water consumption, so it is clear that they could get clear benefits for the solar thermal technology, saving energy from non-renewable sources like gas or electricity produced with coal.

SOLAR THERMAL REGULATIONS AND STANDARDS IN BRAZIL

The experience has shown in several countries that political support has a big impact in the development of renewable energy systems. In order to be successful, the political policies have been created, complying with the following basic requirements:

- Long term policies well developed and implemented.
- Establishment of a clear scope.
- Support activities to the policies:
  - Disseminations policies.

- Training programs
- Financial schemes

The policies are the base for discussion and development of local, regional and national regulations to encourage the use of solar thermal systems in buildings.

Currently in Brazil is not compulsory the installation of solar thermal systems in the whole country. There are regional regulations already developed for the regions of Sao Paulo and Río de Janeiro. Some of those areas also developed local regulations to standardize specific requirements not covered by the regional regulations.

Below there is a list of approved solar thermal regulations in Brazil:

Regional regulations

- Sao Paulo. Lei N° 326 (2007) Installation of solar thermal systems in public buildings [4].
- Rio de Janeiro. Lei N° 5184 (2008) Installation of solar thermal systems in public buildings [5].

Local regulations

- Sao Paulo. SP – Lei 14459. Installation of solar thermal in new buildings [6].
- Sao Paulo. Decreto N° 49.148. Regulates the Lei 14.459 – Solar thermal system [7].
- Sao Paulo, Peruibe. SP-Lei 2883-2007. Installation of solar thermal system in public and private buildings [8].
- Minas Gerais. Lei 31-2006. Lei 31-2006. Installation of solar thermal or gas boilers in hospitals, schools and nurseries [9].

Also at the time of writing there are already several regulations waiting for approval. Among them there are two national regulations that will have an impact on the use of the solar thermal in Brazil:

- Projecto de Lei 478330-2007 – It makes compulsory the use of solar thermal in buildings with financial support from organizations like SBPE, FGTS, FAT or OGU [10].
- Projecto de Lei N° 1563-2007 - Use of renewable energy resources for distributed energy generation in Brazil [11].

Apart from the federal (national), regional and local regulations that regulates the use of solar thermal in different type of buildings, technical standards to standardize the way the systems are sized, dimensioned, installed and maintained.

- ABNT NBR 15569:2008 Solar thermal systems with direct circuit – Project and installation [12].
- ABNT NBR 15747/1:2009 Solar systems and their components – Solar Collectors – Part 1 – General requirements [13]
- ABNT NBR 15747/2:2009 Solar systems and their components – Solar Collectors – Part 2 – Test Methods [14]

Water Consumption in Hotels

The hotel industry is an important sector of the tourism industry with a growth in Brazil of 17% in 2011 over 2010 [15]. This hotel business accounts for a significant amount of resources consumption and environmental impact it generates. Hotel facilities are among the top five in terms of energy consumption in the commercial building sector (minor only to food services, health care, and certain type of offices) [16]. Giving a number, Bothdanowicz estimates that European hotels emit more than 10Mt of carbon dioxide annually [17].

Bothdanowicz studied that the consumption profiles of hotels differ from the other commercial buildings [16]. Factors influencing facility performance can be classified into physical and operational parameters. The physical parameters are common to most buildings like, size, structure and building design. Operational parameters like type of energy and water systems installed, operation and maintenance, depend on each particular building.

The resources used in hotels are influenced also by operational features like operating schemes, catering outlets, laundries, swimming pools and spas, recreational and business centers, services offered and fluctuation in occupancy levels.

Hot water is a big part of a hotel facility´s consumption, which is attributable mainly to the guestrooms, swimming pools, and in kitchen and laundry operations. Table 1 shows benchmark values based on the Brazilian standard NBR 7198/82 [18] and the UK Plumbing Engineering Services Design Guide [19]. Although the Brazilian standard has been superseded by the NBR 7192/93 [18] which does not contain any table for the consumptions, the values are still widely used by engineers and manufactures in Brazil. The last two columns show also the benchmarks for the hot water storage.

Table 1. Benchmark for hot water consumption in hotels, l/guest-night

Room/area	Daily (liters)	Stored (liters)
Hotel room 4/5 star (room, 2 guest)	72 (36 l each)**	45*
Large Spa (swimming pool, per person)	20*	20*
Dinning (Restaurant, per cover)	12**	6*
Meeting space	2*	1*
Breakfast bar (per person)	2*	1*
Canteen Staff	15*	5*
Ballroom (Restaurant, per cover)	12**	6*
Gym (per person)	20*	20*

Source: * [19], ** [18].

SOLAR THERMAL REQUIREMENTS FOR HOTELS

There are at the moment several countries that are making compulsory the use of solar thermal to reduce the energy used for hot water in high consumption buildings like hotels. As one of the most world demanding code, the Spanish "Código técnico de la edificación" [20], in its chapter DB HE, related to energy savings, states the minimum contribution from solar thermal to solar hot water. In Germany, all new built homes built from January 2009 will be required to install renewable heating systems requesting a minimum of 20% of the hot water demand to be provided by a renewable system like solar thermal. There are also other countries like Irelands and Portugal, which although their national standards are not as restricted as the Spanish one, also request the use of solar thermal in their new build homes.

In Brazil, at the moment there is not a regulation for solar thermal requirements for the whole country, however some cities like Sao Paulo have developed regional regulations that require a minimum of solar thermal production for the total annual demand of the building. In Sao Paulo the local regulation Lei 14459 [6] requires that all new building shall provide 40% of the annual demand from solar thermal.

The Brazilian organization Eletrobras, responsible for the energy generation and distribution, is trying to go even further in the renewable requirements, creating the PROCEL certification. The objective of PROCEL, "Programa Nacional de Conservaçao de Energia Electrica" [21] (National Program Electrical Energy Conservation), is to promote the reduction of electrical energy production in the country and rationalization in order to reduce losses and reduce government investments in electrical distributions upgrades. This certification evaluates the energy efficiency rate for commercial and public buildings. The idea in mind is to minimize the electrical consumption in buildings using alternative systems. As hot water can be easily generated by solar thermal systems, PROCEL gives a point if the solar water system produces at least 60% of the annual hot water demand [22] [23]. This point is necessary if the developer is looking for an A-rating in the PROCEL certificate.

Solar Thermal Distribution Systems for High Rise Buildings

There are three main types of solar heating systems used in high-rise buildings: central hot water system, collective distributed hot water system and individually distributed hot water system [2]. Collectors for central heating are normally installed on the roof where sufficient space must be provided.

In high rise buildings, the increase of area split in several floors, a higher consumption demand is associated to a limited building perimeter. In this case the roof may not provide enough space for the amount of solar collectors required. Besides, the top of the building is one of the important features in high rise buildings, so solar collectors will fight for space with restaurants, swimming pools and viewpoints. All those factors are producing a further integration of the solar collectors in the buildings, being located integrated with the facade or in canopies.

There are two main types of solar collectors: non-concentrating or stationary and concentrating [24]. A non-concentrating collector has the same area for intercepting and for absorbing solar radiation, whereas a sun-tracking concentrating solar collector usually have concave reflecting surface to intercept and focus the sun´s beam radiation to a smaller receiving area, thereby increasing the radiation flux. For building applications, although some sun-tracking collectors have been installed in a few roofs, the majority of the solar thermal solutions have been installed with stationary solar collectors.

There are three types of stationary solar collectors: flat plate collectors (FPC), compound parabolic collectors (CPC) and evacuated tube collectors (ETC) [24]. In a flat plate collector the sun passes through a glass and impact in a high absorber plate surface which transfer the energy to a fluid transfer medium in the fluid tubes to be carried away for storage and use. The CPC have the capability of reflecting to the absorber the incident radiation within wide limits. The necessity of moving the concentrator to track the change of orientation is reduced by using two sections of a parabola. Evacuated tube collectors (also known as vacuum tube collectors) have been developed to avoid issues related to the use of flat-plate solar collectors, especially for warm climates where high temperatures, condensation and moisture may cause an early deterioration of internal materials resulting in reduced performance or system failure. For all those reasons vacuum tube solar collectors are more efficient for applications with high temperatures systems. However their geometry marked by the shape of the vacuum tubes, their higher price, and also the appearance, being plat plates collectors more even-looking, makes sometimes the flat plate collectors a more suitable architectural tailored solution.

In terms of hot water distribution for high rise buildings there is always an issue with the pressure at the different levels due to the height of the building. In order to balance the systems there are several solutions available depending on if the generation and the distribution are collective or individual. The Figure 1 shows typical installations for high rise buildings up to 20 levels, in case the collectors are on the top of the roof or at low level. For facade integration systems any of the proposed systems can be used depending on the connection form the collectors to the distribution system

The Figure 1A shows a collective generation and collective distribution system. The hot water is supplied to the different levels where maintenance valves are installed for isolation of the installation. This system is typical when the distribution system has only one owner or when the metering is provided by the landlord. As pressure increases at different heights, pressure valves are required every certain levels to guarantee that an accurate pressure at every level. This system is sensitive to Legionella. The bacteria could grow in the different parts of the system. For this reason maintenance recirculation should be done, with water above 70°C when the system is not in use. The electrical back use will be used when the generation cannot cover the demand.

Figure 1B shows a collective generation and individual distribution. In this case, the generation is circulated in a close circuit avoiding problems with Legionella and pressures. The pressure of the water dropping is compensated with the pressure of the water rising. In this case each final user will provide a heat exchanger and the cold water intake will be supplied from the final user. This solution is better for residential buildings as the bill for the tenant electricity and cold water is already covered by the tenant.

Figure 1a and 1B show two cases. The first case when the solar collectors are installed on the roof and a second case when the solar collectors are installed at a location close to the ground or lower part of the facade. In the case of the low level installation the distribution pipes rise through the building to avoid problems with the pressure to the height of the building.

Figure 1C shows a typical home application that is normally used in high rise buildings in isolated plantrooms for swimming pools applications or penthouse at the top of the building.

In terms of the solar thermal system, currently most of the solar collectors are installed with the main propose of preheating domestic hot water (DHW) or cover part of the space heating demand [25]. However in high rise buildings, especially in buildings with high consumption demand, the applications are mainly to cover a percentage the hot water, otherwise the required area for the solar collectors would be massive. Evacuated tube solar collectors are increasingly in use compares to flat-plate collectors, because they capture the sun better as

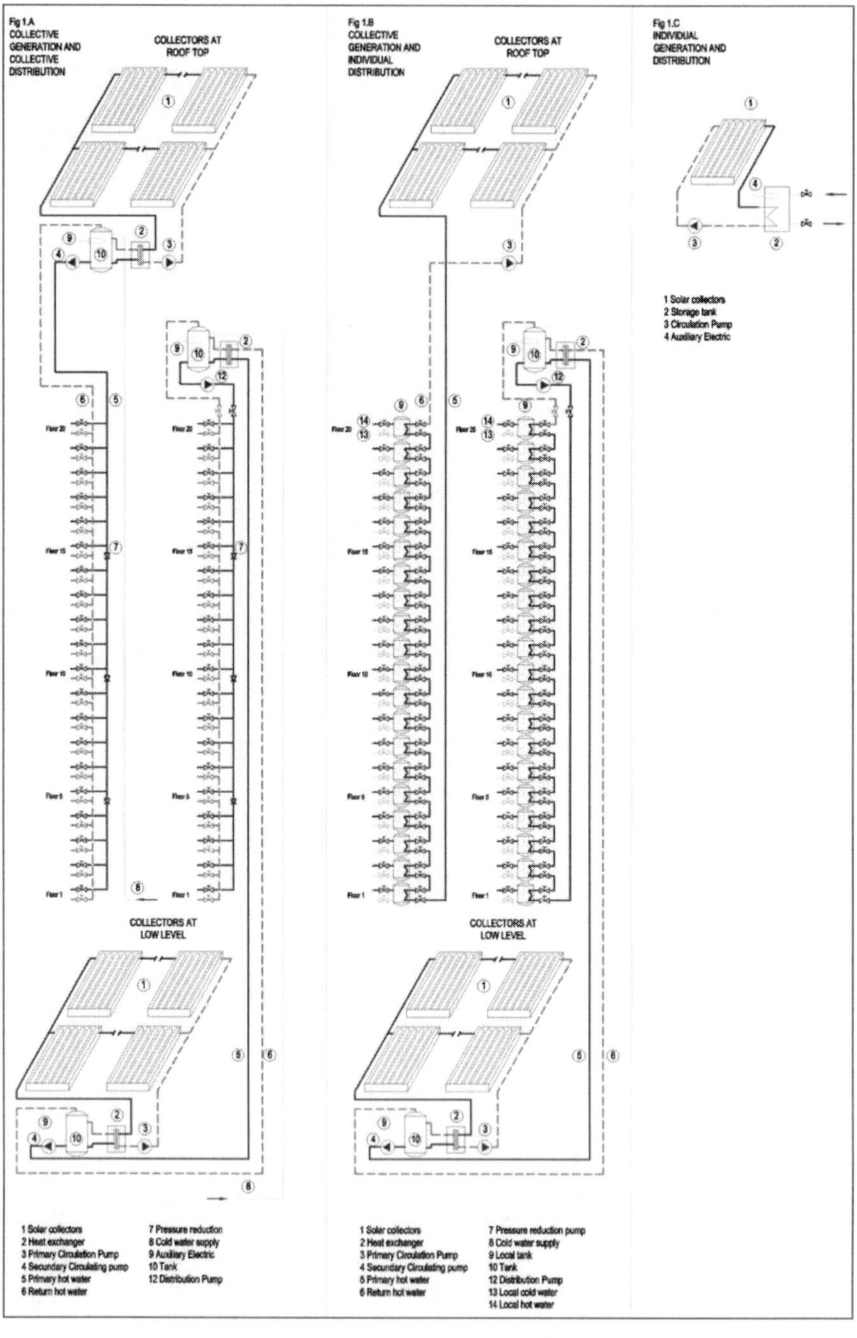

Figure 1. Typical solar thermal installations and distribution in high rise buildings.

they have a greater surface area exposed [26]. An evacuated tube collector operates with thermosyphon flow inside a vacuum- sealed tube, which can minimize the heat transfer though thermal conduction and convection.

For the solar collectors' locations, apart from the typical solution on the roof, some other options have been studied, especially for building renovation market.

Advanced glazed balconies enclose the collector into an openable glass element [25]. The advantage of this system is the easy maintenance in comparison to the facade integrated systems and also that each apartment or hotel room will have a separate system.

Facade collector is becoming the favorite solution for the architects as it can be integrated in convenient not glazed areas. But there are several disadvantages like colors and size to match with the architectural grid and design, the collector is only part of the facade, connection to the roof and other plant connections and facade contractor need to liaise with the solar thermal contractor at early stage are making a difficult coordination at all the stages of the project.

One of the technical challenges in high rise buildings is the pipes vertical distribution due to pressure increase when the distribution is form the roof or pressure drop when the distribution is from the bottom of the building. This issue becomes even more important in the nowadays typical mixed developments, where the available location for the panels is close to ground level. For those mixed developments the basements are built as car parks, the levels around the ground become commercial areas and the higher levels are either residential, offices or a hotel. In this case, if the distribution is from the bottom, normally the space available for plantrooms is just underneath the commercial spaces, that may be even on the second basement (first basement is normally used for supermarkets). In this case pipe will run through several levels that are not serving and the pressure will drop continuously. This issue will make different pressures at different levels, being the lower levels in overpressure to get the pressure required in the areas located at the top of the building. A solution to this problem is to run the main distribution pipe to the top of the building and locate storage tanks in a plantroom.

LEGIONELLA

Legionnaires´ disease is a potential fatal from of pneumonia which can affect anybody, but principally affects those who are susceptible because of age, illness, immunosuppression, smoking, etc. It is caused by the bacterium Legionella pneumonia and related bacteria. Legionella bacteria also cause less serious illnesses which are not fatal or permanently debilitating [27].

Although Legionella is more dangerous in air conditioning systems because the air spreads the bacteria in the ventilated ambient there is also need for hot water systems to include measures to reduce risks from bacteria particularly Legionella. In the UK the BS6700:2006 [28] standard recommends that hot water should be stored at a temperature of not less than 60°C [29], as the organisms will not survive above 60°C.

A possible solution is to use heat exchangers. In this way the storage water is always in a close circuit far from contact from the final user. This application is not effective in general for home applications. However, systems have been developed for home applications like using a residential thermal station (RTS), as it reduces the Legionella bacterium of hot water production, because the hot water in the RTS is produced from fresh cold water using an indirect heat exchanger [30].

As the optimum water temperature for hot water use, normally hot water storage at this temperature, the system is susceptible to grow Legionella, as the optimum growth temperature is between 35°C to 46°C, being instantaneously killed at temperatures higher than 70°C. At temperatures higher than 50°C, the bacteria can survive up to 5 to 6 hours [31]. This allows a way to program automatic maintenance solutions to ensure that Legionella does not exist in hot water close circuits. A temperature boost is produced by the electrical support resistances located in the water storage tank to raise the temperature in the tank above 70°C overnight. Extreme care should be taken to avoid supplying hot water from the tank during boost periods. This is one of the reasons, apart from resilience, why in high consumptions buildings two storage tanks are installed.

CASE STUDY: 5 STAR HOTEL IN SAO PAULO CITY CENTER

According to the study done by Danish consulate in Brazil, there are 5.2 million of international tourists in Brazil, which is still below the potential capacity [32]. Along the economic growth, Brazil has experienced a large expansion of their middle class which also raise the national tourist rate. If we add to that the World Cup in 2014 and the Olympics in 2016, the current cruise and coast line development and the cheap loans available for the hotel sector, is very clear that many hotels will be built in the next years in Brazil. As an example of those hotel developments the following case study propose a solution for a high rise hotel in the city center of Sao Paulo, analyzing the requirements for solar thermal and proposing solutions for the technical issues for those particular buildings.

The proposed hotel is part of a prospective mixed development in the city center of Sao Paulo. Figure 2 shows a schematic representation of the development, showing the uses of the different levels. This development will be divided in two main assets, a commercial part with retail units and a car park from levels Basement 5 to level 02, and a hotel from level 03 to level 24.

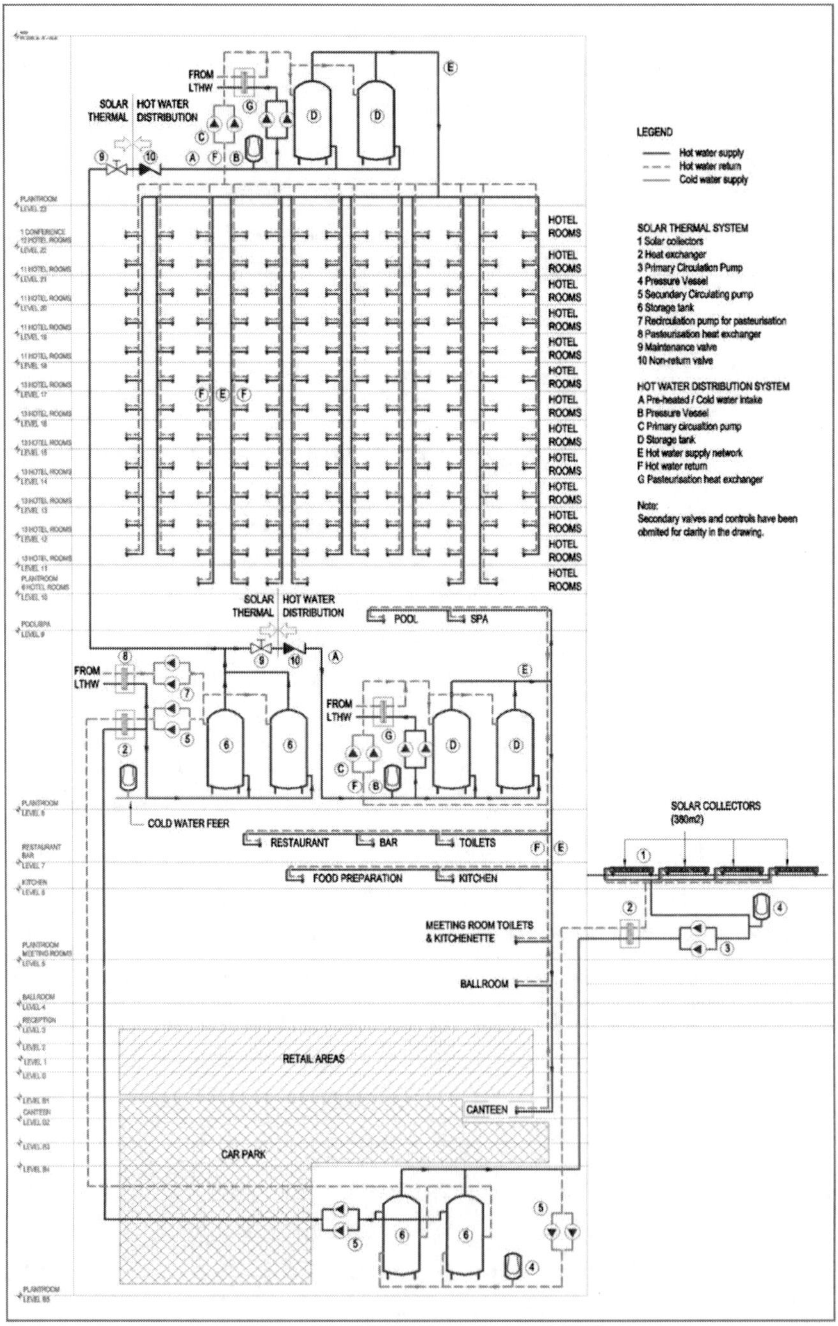

Figure 2. Hotel schematic and hot water distribution layout.

The hotel is divided in two main areas. From level 03 to 09 the hotel areas will provide hotel services for guests and visitors, like the restaurant, bar, ballroom, meeting rooms, gym and spa. The upper levels, 10 to 22, will provide space for guest rooms, with a special conference area at level 22.

Although the hotel is part of the development, it has been designed as a separate asset as it will be let or lease to a hotel operator which will run it. With this concept in mind, the services in the development have been provided separately for the commercial areas and the hotel. This is the reason that although, as a part of a new development, the commercial part requires solar thermal production, it will not be included in this study which is focus in solar thermal for high rise buildings with high density population.

For water installations for high rise buildings there is always a challenge with loss of pressure due to the height of the building. For this reason, intermediate plantrooms have been included in the hotel design to accommodate water storage and pumps. In developments located in expensive areas like the city center of Sao Paulo, the useful areas, especially above the ground, need to be maximized for public use. In these cases, sometimes plant areas are not in the optimum technical location, but in a more suitable location for the developer interests. Figure 2 shows the locations for the intermediate plantrooms at level 8 and level 23 in convenient locations for the water distribution. However the first storage plantroom, located in basement 5, would be better located close to the solar collectors at level 06 or below. Unfortunately, space was not available at that area.

Other areas, like the staff changing rooms and hotel staff canteen, level Basement 02, have been also moved out of the public areas. This means that hot water distribution will cross in a riser the commercial areas to supply hot water to the isolated areas.

Table 2 shows the calculation for the hot water consumption in the hotel for a full day occupation. Those values will give daily water demand and storage required to provide a good service in the hotel. The water densities for the different areas are the ones shown previously in Table 1.

Table 2. Daily hot water demand and storage

Level	Use	No People/ room	daily (liters)	Total daily (liters)	Storage (liters)	Total storage (liters)	Reduced storage (l) - preheat (+)
22	Conference	12	2*	24	1*	12	6
	Rooms	7	72**	504	45*	315	157,5
21	Rooms	11	72**	792	45*	495	247,5
20	Rooms	11	72**	792	45*	495	247,5
19	Rooms	11	72**	792	45*	495	247,5
18	Rooms	11	72**	792	45*	495	247,5
17	Rooms	13	72**	936	45*	585	292,5
16	Rooms	13	72**	936	45*	585	292,5
15	Rooms	13	72**	936	45*	585	292,5
14	Rooms	13	72**	936	45*	585	292,5
13	Rooms	13	72**	936	45*	585	292,5
12	Rooms	13	72**	936	45*	585	292,5
11	Rooms	13	72**	936	45*	585	292,5
10	Rooms	6	72**	432	45*	270	135
		Total hotel rooms		**10680**		**6672**	**3336**
9	Pool/spa	40	20*	800	20*	800	400
8	Gym	30	20*	600	20*	600	300
7	Restaurant	120	12**	1440	6*	720	360
	Bar	50	2*	100	1*	50	25
6	Kitchen	Included					0
5	Meeting	54	2*	108	1*	54	27
4	Ballroom	240	12**	2880	6*	1440	720
3	Lobby	Included					0
B2	Canteen	50	15*	750	5*	250	125
		Total hotel services		**6678**		**3914**	**1957**
		Total hotel		**17358**		**10586**	**5293**

(+) Using 2 reheat periods the storage is reduced to half of the capacity as the temperature of the cold water from the main rises as soon as it enters the tank.

Source: * [19], ** [18]).

For this large amount of hot water demand, the solar thermal system will have to be oversized to provide the required conditions during the winter time, and the system will be overheated during the summer. For this reason, the solar thermal system will work as a preheating system to the hot water distribution system. In absence of enough energy from the solar thermal system, an electric heater will heat the water to the required temperature.

The advantage of the preheat system is that it reduces the storage capacity to half, minimizing the storage space. As it is shown in Figure 1, the storage for the hot water is divided between the roof plantroom which provides hot water to the hotel rooms and the plantroom at level 8 which provides hot water storage for the hotel services levels. Table 3 shows the cylinders selected for the storage on each plantroom, shown as "D" in Figure 2

Once the daily demand is calculated, a calculation can be done for the percentage of this hot water that needs to be generated by solar thermal. The installation will be sized to comply with both, the 40% minimum requirement from by "Lei 14459" for Sao Paulo, but also the 60% required by PROCEL to achieve the A-rating. To ensure that the system will be always above the required 60%, taking into account small variations in the annual weather performance, the calculation assumes a 5% margin, in order to guarantee a minimum of 65% of the energy production from hot water from the solar thermal system. To achieve the 65% of the total year demand, several types of solar collectors were compared in the selection being the solar vacuum solar tube collectors the ones that show a better performance for large installations, especially for warm countries like Brazil. The solar vacuum panel Vitosol 300T-SP3 ($3m^2$) [33] was selected due to the cost-efficiency rate and its low heat loss coefficient.

There are in the market several types of software to calculate the number of panels, energy performance and payback periods. Among them, RETscreen [34] is one of the best renown in the world due to their large weather and manufacturer's data base. This tool allows the user to make calculations for a certain location like Sao Paulo, with a commercial selected panel.

Table 3. Calorifiers selection

Areas served	Plant location	N° tanks	Capacity	Total
Hotel rooms	Level 23	2	1800	3600
Hotel services	Level 08	2	1500	3000

Table 4. Comparative calculations versus RETScreen

Item	Calculated	RETScreen
Number of rooms	160	160
Daily hot water use (L)	17358	12128
Minimum intake water temperature (°C)	10	18
Maximum intake water temperature (°C)	18	21.2
Number of collectors	72	56

Using as an input the required daily hot water and storage for the hotel, RETscreen calculates the estimated hot water use according to the software database. Table 4 below shows the numbers for the different items. Where numbers from RETScreen are lower than calculated numbers, the calculated number was introduced in the software superseding the RETScreen value. In example, for the daily hot water consumption, RETScreen does not take into consideration the high water demand for hotel services for this application, so the RETScreen value of 12128 l is superseded by 17358 l introduced as an input.

Using 72 solar collectors the calculated area is approximately 310 m^2. Due to architectural constrains at the top of the building, the provided roof had not enough area to allocate the panels. A solution to integrate the panels was to increase the size of the main entrance canopy, above the hotel entrance, to an area of 400 m^2 to accommodate the solar collectors. A study was carried out to check that there were no shadows due to other existing or prospective buildings in the area. The solar thermal system will work preheating storage tanks (6 in Figure 2) within temperatures from 10°C (temperature of the cold water in the worst condition), to 45°C which is the maximum temperature at the tab.

The building management system will control the temperature in designated positions to keep the temperature level in the tank high enough to guarantee 45°C in the closest tab. The control will control the temperature through predictive control [35] to minimize the use of the electrical back-up, switching it on only when is strictly necessary.

As mentioned in the calculations, the system uses two reheat periods in order to minimize the storage space. This solution makes the system slightly more inefficient compared to a system with a larger storage, but makes it more cost effective, reducing the storage and therefore minimizing the capital cost. Taking this into account, RETScreen compares the proposed system to an electrical boiler. Table 5 shows the electrical consumption and electricity costs with and without the solar thermal system.

Table 5. RETScreen calculation with or without solar thermal system

Item	Electrical boiler only	Electrical boiler + solar thermal
Fuel type	Electricity	160
Efficiency	90%	90%
Annual electricity consumption (MWh)	254,8	87,6
Electricity rate ($)	0,07(*)	0,07(*)
Annual electricity cost ($)	17835	6131

(*) Building assumed to be supplied in High voltage by Electropaulo using blue tariff. Water heating to be supplied from the electrical network at off-peak times (form 8:30 pm to 5:30 pm). If electrical supply is required at peak time, from 5:30 to 8:30 pm, a tariff of 0,12$ will be applied to the electricity rate.

As it can be seen from Table 5, there is a saving of 65% (1-(6131/17835)/100) in the electricity cost using solar thermal for hot water.

Although the main reason of providing solar thermal hot water was due to the regulations making it compulsory, there is a clear saving for the hotel operator in running cost. This is making that in areas where the solar hot water is not available; some hotel operators are requesting the system in their buildings. To analyze the payback period of the investment there are not many studies of the capital cost of solar thermal water for hotel applications. Erwin Edwards presented a study [36] for the capital cost according to the number of rooms for a hotel. In this study, he declares a cost of 80.000$ for a 200 rooms hotel, however is unlikely that in this assessment was including a preheating system, reason why 100.000$ is taken to calculate the payback period using RETScreen. With this assumption, the payback period for the solar thermal installation proposed is around 7.2 years for an equity payback and around 9.4 years for a simple payback. This payback period is reasonable even for countries like Brazil, where there are not available at the moment grants for solar thermal systems.

There are already installations of solar thermal that have been successfully working for more than 25 years. In a case study, Todorovic [37] studies the efficiency of a combined solar thermal and seawater heat pump system in the production of cooling and hot water for a hotel of 754 rooms and 220 apartments. The presented test on the installation over 20 years of operation concludes that the solar thermal systems have been working successfully with minor maintenance, performing with up to 75% efficiency during the last 10 years. The combined system had a simple payback period of only 3 years, justifying the use of renewable energies in the production of hot water and cooling. However the use of combined renewable systems is related to availability. In our case in Sao Paulo there is not availability of sea water or similar source and the expensive price of land in the center of Sao Paulo will challenge the area required for the seawater heat pump plantroom.

Table 6 shows a comparative [37] [38] of some hotels in Europe that are currently using solar thermal energy for hot water and heating, showing the justification of the solar thermal solution even in areas where grants or subsidies are not available.

Table 6. Other built case studies

Hotel	Year	Beds (No)	Collectors area (m^2)	Integration	Energy reduction kWh/a (%)	Replaced energy source
Slovenia Beach (Budva, Montenegro) (*)	1983	974	2500	Roof car park	3.660.000	Seawater heat pump
Fortuna (Germany) (**)	2004	70	53	Roof	34.000 (7.7%)	Heating Oil
Accor - Novotel Sophia Antipolis. Valbonne (France) (**)	1999	97	113	Roof	94.650 (48%)	Electricity
Consortio Parcatinat (Italy) (**)	2004	200	150	Roof	180.000 (55%)	Natural Gas
Die Fabrik (Berlin, Germany) (**)	1999	23	27	Roof	35.000 (7.1%)	Natural Gas
Dom paraplegiokov (Slovenia) (**)	2006	120	78	Roof	Not available	Liquid gas
Hotel du golf de Valexcure (France) (**)	2003	40	90	Roof	68.000	Electricity

(Source: *[37] y **[38]).

To minimize the risk of Legionella, the system is calculated to be at least 2 hours at 65°C. A water intake will be provided from the heating system, through a pasteurization heat exchanger, in case the temperature in the tank did not achieve 65°C during 5 continuous hours during the week, to effectively remove any Legionella from the system. The temperature boost will be done during maintenance hours when nobody is using the system. The use of two storage tanks provides service resilience in case of maintenance or failure. A similar system, with a heat exchanger (G in Figure 2) from the heating system, is provided to the distribution storage tanks (D in Figure 2). This is especially important on those tanks, as hot water temperature must not be higher than 45°C at the taps. Electric heaters are provided in the storage tanks to do the pasteurization when the heating is not working, and as a back up to achieve the hot water design temperature when the solar energy is not available.

Further to the requirements from the regulations, there was a proposal to size up the system to cover the 100% of the hot water demand during the winter, using the heating surplus to produce cooling with a triple state LiCl-H_2O [39]. Unfortunately this option was dismissed as it is not required neither by the regulations nor by PROCEL.

CONCLUSION

The proposed solutions for solar hot water systems for high rise building applications minimize the issues with the pressure distribution due to the height of the building. The preheat system shown in the case study provides the required resilience to the hot water system for 24 hours operational buildings like hotels. In this way, the electrical system can support the whole installation in the meantime the solar thermal system is maintained. This segregation is also beneficial for Legionella treatment. The solution of boosting the temperature in the tank, by electric heaters or using the water from space heating will minimize the risk of Legionella in the hot water system.

Regulations like "Lei 14459" for the city of Sao Paulo, which requires the 40% of the hot water to be provided by solar thermal systems, in the same way than environmental certifications for buildings like PROCEL, which requires 60%, are pushing the use of solar thermal technologies for new buildings, being the systems not only applied to residential buildings, but also to commercial buildings with a high hot water consumption like hotels.

The case study for a 5 star hotel shows the large amount of hot water required for the operation of the hotel every day. To comply with the requirements from the regulations and PROCEL, a large number of panels are required. As the architects are using the top of the building as one of the building's feature, the proposed solution locates the solar collectors above the hotel canopy above the hotel entrance to integrate the collectors to the building without affecting the top of the building,

Using RETScreen, the calculation justifies the production of up to 65% of the solar thermal meeting the 60% and 40% required by PROCEL and Sao Paulo regulations respectively. This calculation also justifies the use of the system even when the regulation does not require it. As shown in the figures, a payback of around 8 year justify the savings in electricity or fuel consumption, making the proposed solar thermal system optimum for high consumption buildings like hotels.

References

[1] *The history of solar*. U.S. Department of energy. Energy efficiency and renewable energy.

[2] Jie, Shi; Wei, Su; Mingya, Zhu; Huaning, Chen; Yiqun, Pan; Shi, Wan; Yawei, Wang. Solar water heating system integrated design in high-rise apartment in China. *Energy and Buildings*, 58, (2013), 19-26.

[3] Ricardo, Luiz Tows; Cesar, Miranda Mendes. O estudo da verticalização urbana como objeto da geografia: Enfoques e perspectivas metodológicas [Study of urban buildings as object of geography: Approaches and methodological perspectives]. Simpósio de Estudos Urbanos. *Agosto*, 2011.

[4] Lei N° 326, de 2007 - *Dispõe sobre a instalação de sistema de aquecimento de água por energia solar em edificações de natureza pública, no âmbito do Estado de São Paulo* [*Installation of solar water heating system in buildings of public nature, in the scope of the State of São Paulo*].

[5] Lei N° 5184, de 2008 - *Dispõe sobre a instalação de aquecimento solar de água em prédios públicos no Estado do Rio de Janeiro* [*Installation of solar water heating in public buildings in the State of Rio de Janeiro*].

[6] São Paulo/SP - Lei 14459 - *Dispõe sobre a instalação de sistema de aquecimento de água por energia solar nas novas edificações do Município de São Paulo* [*Installation of water heating system by solar energy in the new buildings of the Municipality of São Paulo*].

[7] São Paulo/SP - Decreto N° 49.148 - Regulamenta a Lei 14.459 - *Aquecimento de água por energia solar* [São Paulo / SP - Decree No. 49,148 - Regulates Law 14,459 - *Water heating by solar energy*].

[8] Peruíbe-SP -Lei 2883-2007 - *Dispõe sobre a instalação de sistema de aquecimento solar de água em edificações públicas ou privadas e dá outras providências* [*Installation of solar water heating system in public or private buildings and other measures*].

[9] Lei 31-2006 - *Dispõe sobre a implantação e instalação de aquecedores solares e/ou gás natural nas residências, hospitais, escolas e creches* [*Installation of solar heaters and / or natural gas in homes, hospitals, schools and day care centers*].

[10] Projeto de Lei 478330- 2007 - *Cria a obrigatoriedade de utilização como fonte subsidiária de energia, sistema de aquecimento solar de água em imóveisfinanciados com recursos do SBPE, FGTS, FAT e do OGU – orçamento geral da união e dá outras providências* [*Mandatory use as a subsidiary source of energy, solar water heating system in properties financed with funds from SBPE, FGTS, FAT and OGU - general budget of the union and other measures*].

[11] Projeto de Lei N° 1.563, de 2007 *(Do Sr. Paulo Teixeira, Nilson Pinto, Bernardo Ariston e outros): Dispõe sobre fontes renováveis de energia, com o objetivo de promover a universalização, a geração distribuída e a racionalização energética, e altera a Lei*

n° 10.438, de 26 de abril de 2002, para modificar o Proinfa e aumentar a participação de fontes alternativas na matriz energética nacional [*Provides for renewable energy sources, with the objective of promoting universalization, distributed generation and energy rationalization, and amending Law No. 10,438, of April 26, 2002, to modify Proinfa and increase the participation of alternative sources in the national energy matrix*].

[12] ABNT NBR 15569:2008. *Sistemas de Aquecimento Solar en Agua en Circuito Directo – Projeto e Instalacao* [*Solar Water Heating Systems in Direct Circuit - Design and Installation*]

[13] ABNT NBR 15747/1:2009 Sistemas Solares e seus Componentes – Colectores Solares – Parte 1 – Requisitos Gerais [*Solar Systems and their Components - Solar Collectors - Part 1 - General Requirements*].

[14] ABNT NBR 15747/1:2009 *Sistemas Solares e seus Componentes – Colectores Solares – Parte 2 – Métodos de Ensaio* [*Solar Systems and their Components - Solar Collectors - Part 2 - Test Methods*].

[15] Hoteleria en Números. Brasil 2012. *Forum the operadores hosteleiros do Brasil* [Hotes in Numbers. Brazil 2012. *Forum for hotel operators in Brazil.*]

[16] Paulina, Bohdanowicz; Ivo, Martinac. Determinants and benchmaking of resource consumption in hotels – Case of study of Hilton and Scandic in Europe. *Energy and buildings*, 39, (2007), 82-95.

[17] Bohdaniwicz, P. European hoteliers ´environmental attitudes: greening the business, *Cornell Hotel and restaurant quarterly*, 46, (2005), 188-204.

[18] ABNT NBR 7198:1982 Projeto e execução de instalações prediais de água quente [Design and execution of hot water building installations]. *Superseded by ABTN*, 7198, 1993, Projeto e execução de instalações prediais de água quente.

[19] The institute of Plumbing. *Plumbing engineering Services design Guide*. Essex, UK., 2002.

[20] Código Técnico de La Edificación [Technical Building Code]. *RB*, 1371/2007, 19 October, 2007.

[21] PROCEL. http://www.eletrobras.com/elb/procel/main.asp. *Programa Nacional de consevaçao de energia elétrica* [*National Program of conservation of electric energy*].

[22] PROCEL. RTQ-C. *Regulamento Técnico da Qualidade para o Nível de Eficiência Energética em Edificações Residenciais* [*Quality Technical Regulation for the Energy Efficiency Level in Residential Buildings*].

[23] PROCEL. RAC-C. *Regulamento de Avaliação da Conformidade do Nível de Eficiência Energética de Edifícios Comerciais, de Serviços e Públicos* [*Regulation of Conformity Assessment of the Energy Efficiency Level of Commercial Buildings, Services and Public Buildings*].

[24] Soteris, A. Kalogirou. Solar thermal collectors and applications. *Progrees in energy and combustion science*, 30, (2004), 231-295.

[25] Kasten, Voss. Solar energy in building renovation – results and experience of international demonstration buildings. *Energy and Buildings*, 32, (2000), 291-302.

[26] Tin-Tai, Chown; Zhaoting, Dong; Lok-Shun, chan; Kwong-Fai, Yu Bai. Performance evaluation of evacuated tube solar domestic hot water systems in Hong Kong. *Energy and Buildings*, 43, (2011), 3467-3474.

[27] Health and safety executive, Legionnaire's Disease – *The Control of Legionella Bacteria in Hot Water*, 2000. http://www.hse.gov.uk/pubns/books/18.htm.

[28] British standards institution, BS 6700:2006 + A1:2009 *Design, installation, Testing and maintenance of Services supplying Water for Domestic within Buildings and their Curtilages – specification*, BSI, London, 2009.

[29] Boait, PJ; Dixon, D; Fan, D; Stafford, A. Production efficiency of hot water for domestic use. *Energy and buildings*, 54, (2012), 160-168.

[30] Tomasz, Cholewa; Alicja, Suita-Olcha. Experimental investigations of a decentralized system for heating and hot water generation in a residential building. *Energy and buildings*, 42, (2010), 183 - 188.

[31] Manual de Capacitação em Projetos de sistemas de Aquecimento Solar [Training Manual on Solar Heating System Projects]. *Abrava.*, 2008.

[32] The Brazilian Hotel Sector. *The Danish consulate general and the embassy of Denmark*. Sao Paulo, 2010.

[33] Vitosol 300-T. Technical manual. *Viessmann.*, 02/2011.

[34] RETScreen 4, *Clean Energy Project Analysis Software.* www.netscreen.net. Canmet ENERGY. NRCan.

[35] Solutions to Reduce Energy Consumption in the Management of Large Buildings. *Energy and Buildings*, Volume 56, January 2013, Pages 66-77. Antonio Colmenar-Santos, Lya Noemí Terán de Lober, David Borge-Diez, Manuel-Alonso Castro-Gil.

[36] Erwin, Edwards; Bernard, Linton. *Comparative Hot Water Systems for Hotels*. Caribean community Secretariat.

[37] Marija, S. Todoravic; Slobodan, Pejkovic; Vido, Zenovic. 3.5MW Seawater heat pump assisted multipurpose solar system's 25 years of operation. *ASHRAE Transactions*, 2010, Vol. 116, part1.

[38] *Solar heating for large buildings.* Solarge. Inteligent building. Europe.

[39] Exergy efficiency analysis in buildings climatized with LiCl–H2O Solar cooling systems that use swimming pools as heat sinks. *Energy and Buildings*, Volume 43, Issue 11, November 2011, Pages 3161-3172 D. Borge, A. Colmenar, M. Castro, S. Martín, E. Sancristobal.

About the Editors

Antonio Colmenar Santos, PhD
Full Professor
Departamento de Ingeniería Eléctrica, Electrónica, Control, Telemática y Química Aplicada a la Ingeniería,
Universidad Nacional de Educación A Distancia, Madrid, Spain
Email: acolmenar@ieec.uned.es

Dr. Colmenar-Santos has been a senior lecturer in the field of Electrical Engineering at the Department of Electrical, Electronic and Control Engineering at the National Distance Education University (UNED) since June 2014. Dr. Colmenar-Santos was an adjunct lecturer at both the Department of Electronic Technology at the University of Alcalá and at the Department of Electric, Electronic and Control Engineering at UNED. He has also worked as a consultant for the INTECNA project (Nicaragua). He has been part of the Spanish section of the International Solar Energy Society (ISES) and of the Association for the Advancement of Computing in Education (AACE), working in a number of projects related to renewable energies and multimedia systems applied to teaching. He was the coordinator of both the virtualisation and telematic Services at ETSII-UNED, and deputy head teacher and the head of the Department of Electrical, Electronics and Control Engineering at UNED. He is the author of more than 50 papers published in respected journals (http://goo.gl/YqvYLk) and has participated in more than 100 national and international conferences.

Enrique Rosales Asensio, PhD
Researcher
Departamento de Física
Universidad de La Laguna, La Laguna, Spain
Email: erosalea@ull.edu.es

Dr. Rosales-Asensio, PhD, is an industrial engineer with postgraduate degrees in electrical engineering, business administration, and quality, health, safety and environment management systems. Currently, he is a senior researcher at the University of La Laguna, where he is involved in water desalination project in which the resulting surplus electricity and water would be sold. He has also worked as a plant engineer for a company that focuses on the design, development and manufacture of waste-heat-recovery technology for large reciprocating engines; and as a project manager in a world-leading research centre.

David Borge Diez, PhD
Lecturer
Departamento de Ingeniería Eléctrica y de Sistemas y Automática
Universidad de León, León, Spain
Email: dbord@unileon.es

Dr. David Borge-Diez has a PhD in Industrial Engineering and an M.Sc. in Industrial Engineering, both from the School of Industrial Engineering at the National Distance Education University (UNED). He is currently a lecturer and researcher at the Department of Electrical, Systems and Control Engineering at the University of León, Spain. He has been involved in many national and international research projects investigating energy efficiency and renewable energies. He has also worked in Spanish and international engineering companies in the field of energy efficiency and renewable energy for over eight years. He has authored more than 25 publications in international peer-reviewed research journals and participated in numerous international conferences.

INDEX

A

actuators, 58, 59
adsorption, 44, 52, 57, 74
adsorption coefficient of performance (COP), 2, 14, 31, 44, 63, 70
air pollutants, 112
air quality, 7, 12, 16, 39, 40, 79, 86, 88, 91, 103, 110
air temperature, 11, 54, 68, 87, 101, 103, 104, 106
ambient air, 10, 66, 81
ambient air temperature, 66
American Society of Heating, Refrigerating and Air-Conditioning Engineers (ASHRAE), 2, 11, 35, 36, 37, 38, 78, 86, 108, 109, 141
automation, 44, 47, 55, 56, 58, 59, 60

B

bacteria, 123, 126
benefits, 7, 18, 34, 40, 112, 116
boilers, 69, 117
building code, 19
building management system (BMS), 44, 46, 58, 59, 62, 65, 66, 133

C

camps of internally displaced people (CIDPs), 78, 79, 80
carbon dioxide, 86, 118
cities, 18, 34, 108, 114, 115, 116, 120
climate change, 18, 39
climate(s), 11, 12, 18, 20, 35, 37, 39, 40, 52, 77, 81, 107, 109, 111, 122
coatings, 39, 95, 99, 109, 110, 111
coefficient of performance (COP), 2, 14, 44
computational fluid dynamics (CFD), 2, 6, 7, 9, 10, 11, 20, 21, 22, 33, 78, 84, 88, 96, 105, 106
condensation, 24, 122
conditioning, 2, 5, 19, 39, 126
conductivity, 23, 79, 93, 96, 101
construction, 4, 14, 17, 23, 45, 50, 55, 60, 63, 80, 84, 90, 91, 92, 104, 107, 114, 115, 116
cool roof (CR), 2, 5, 17, 18, 19, 20, 23, 24, 25, 28, 31, 32, 34, 35, 38, 40, 41, 77, 78,

84, 93, 94, 95, 96, 97, 99, 101, 103, 104, 106, 108, 110, 111, 112
cooling, 1, 5, 12, 14, 18, 20, 22, 23, 24, 25, 26, 27, 28, 30, 31, 32, 33, 34, 35, 38, 41, 45, 46, 52, 53, 54, 56, 57, 59, 62, 64, 66, 67, 69, 70, 71, 72, 73, 77, 79, 81, 83, 86, 91, 92, 93, 94, 95, 99, 107, 109, 110, 111, 134, 135, 141

D

database, 37, 65, 70, 108, 132
direct digital control (DDC), 44, 58, 59
direct expansion (DX), 2, 5, 13, 16, 23, 24, 25, 29, 31, 33
discomfort, 18, 102, 107
distribution, 7, 11, 14, 15, 21, 22, 23, 24, 25, 26, 28, 29, 30, 31, 32, 33, 34, 46, 52, 53, 59, 84, 88, 99, 114, 120, 122, 123, 124, 125, 128, 129, 131, 135, 136

E

electricity, 5, 13, 17, 19, 33, 34, 45, 49, 58, 69, 84, 89, 91, 116, 123, 133, 135, 137
emission, 45, 112
energy, 1, 2, 3, 4, 5, 6, 7, 8, 9, 13, 14, 15, 17, 18, 19, 20, 22, 23, 25, 31, 33, 34, 38, 39, 40, 41, 43, 44, 45, 46, 48, 49, 50, 52, 55, 56, 57, 59, 60, 61, 62, 65, 66, 67, 69, 70, 71, 72, 73, 75, 76, 77, 78, 79, 81, 84, 87, 89, 91, 107, 108, 109, 110, 111, 112, 114, 116, 118, 119, 120, 122, 131, 135, 137, 138, 139, 140
energy consumption, 1, 2, 5, 7, 13, 14, 15, 18, 20, 22, 23, 33, 45, 46, 47, 55, 61, 65, 69, 70, 71, 72, 73, 75, 79, 108, 118
energy efficiency, 3, 6, 13, 14, 31, 45, 60, 73, 120
energy supply, 45, 79, 84
environment, 11, 45, 46, 61, 71, 86, 87, 95, 110, 111
environmental impact, 31, 118
environmental management, 41
environmental variables, 10
equipment, 19, 20, 34, 47, 49, 52, 55, 107, 115
error detection, 71
evacuated tube collectors, 122
evacuation, 96, 106
evapotranspiration, 40
expanded polystyrene insulation (EPS), 2, 20, 24, 78, 84, 93, 97, 101, 108

F

fault detection, 62, 69, 75
fault diagnosis, 69
fluid, 2, 7, 10, 122
fuel consumption, 34, 137

G

greenhouse gas, 34, 41, 111, 112
greenhouse gas emissions, 41, 111
growth, 86, 115, 118, 126
growth temperature, 126

H

heat loss, 56, 81, 131
heat transfer, 3, 12, 18, 50, 56, 87, 95, 125
hemisphere, 80
high rise buildings, 114, 115, 116, 121, 122, 123, 124, 125, 129
humidity, 2, 14, 86, 101, 102, 105

I

immunosuppression, 126

infrastructure, 5, 7, 80
insulation, 2, 5, 20, 49, 84, 97, 108
integrated solar thermal, 114
integration, 2, 58, 113, 115, 121, 122
internally displaced, 4, 5
international development, 2
  development cooperation, 78

L

Legionella, 114, 123, 126, 135, 136, 140
life cycle, 39, 60, 73, 76, 112
living conditions, 4, 7, 48, 78, 79, 80, 103, 108
low cost housing, 2, 78

M

mass, 44, 62, 65, 71
material surface, 94
materials, 6, 14, 17, 20, 23, 25, 38, 40, 45, 60, 70, 79, 88, 91, 92, 95, 96, 99, 107, 122
metropolitan areas, 40, 110
models, 6, 9, 10, 23, 44, 61, 68
moisture, 122
mold, 86

N

natural disaster, 4, 80
natural gas, 138

O

optimization method, 71
oxygen, 46
ozone, 39, 40, 48

P

partial differential equations, 8
pasteurization, 135
payback period, 131, 134
PDEs, 2, 8
physical characteristics, 17, 92, 93
physical structure, 58
plants, 62
pneumonia, 126
pollutants, 86
pollution, 19
polystyrene, 2, 20
pools, 35, 72, 119, 121, 123, 141
population, 4, 11, 129
Portugal, 120
predictive control primary energy simulation, 44
principles, 63
probability, 19
PROCEL, 114, 120, 131, 136, 137, 140
project, 19, 44, 46, 61, 101, 125
proliferation, 15, 45
protection, 4, 49, 50, 56, 58, 59
prototype, 104, 108
pumps, 115, 129

R

radiation, 18, 36, 39, 51, 95, 98, 110, 121, 122
refugees, 3, 4, 5, 34, 35
regression equation, 86
regulations, 43, 114, 115, 117, 118, 120, 134, 135, 136, 137
regulatory requirements, 114
renewable energy, 2, 7, 13, 14, 19, 29, 33, 34, 41, 48, 80, 81, 113, 116, 118, 137, 138
resilience, 127, 135, 136
resistance, 3, 17, 92

resources, 4, 17, 38, 39, 45, 48, 84, 89, 91, 114, 118, 119
respiratory problems, 86
room temperature, 46, 64, 68

S

shelter, 6, 7, 12, 15, 17, 18, 20, 25, 33, 34, 79, 80, 84, 88, 89, 95, 103, 108
solar collectors, 14, 28, 114, 115, 121, 122, 123, 125, 129, 131, 132, 136
solar hot water, 114, 120, 134, 136
solar thermal energy, 5, 14, 134
solution, 5, 6, 17, 18, 20, 23, 32, 33, 34, 65, 77, 80, 81, 84, 90, 91, 92, 93, 95, 96, 99, 101, 102, 104, 106, 107, 113, 114, 122, 123, 125, 126, 127, 132, 133, 135, 136
structure, 48, 49, 66, 103, 119

T

temperature, 3, 7, 11, 12, 13, 14, 16, 18, 24, 37, 38, 52, 54, 57, 63, 64, 66, 68, 69, 79, 81, 84, 86, 87, 94, 95, 98, 101, 102, 103, 104, 105, 109, 112, 126, 130, 131, 132, 133, 135, 136
thermal activation, 44, 52, 57
thermal activation of the components (TAC), 44
thermal analysis, 103
thermal comfort, 2, 5, 7, 8, 10, 11, 18, 19, 20, 31, 33, 34, 37, 38, 48, 77, 78, 79, 86, 87, 88, 92, 93, 101, 102, 104, 107, 108, 109, 111
thermal energy, 14, 57
thermal piles, 44
thermal resistance, 87, 93
transmission, 3, 44, 49, 50, 58

V

ventilation, 2, 3, 5, 12, 15, 16, 20, 22, 23, 25, 34, 35, 37, 54, 56, 59, 63, 65, 68, 77, 81, 84, 86, 88, 91, 92, 93, 96, 99, 101, 103, 104, 106, 108

W

water, 3, 11, 17, 23, 28, 52, 62, 67, 79, 81, 87, 89, 96, 113, 114, 115, 116, 119, 120, 121, 122, 123, 126, 128, 129, 130, 131, 132, 133, 134, 135, 136, 137, 138, 139, 140
water absorption, 79
water evaporation, 81
water heater, 114, 115
wind power, 59

Electric Power: Developments and Issues for Congress

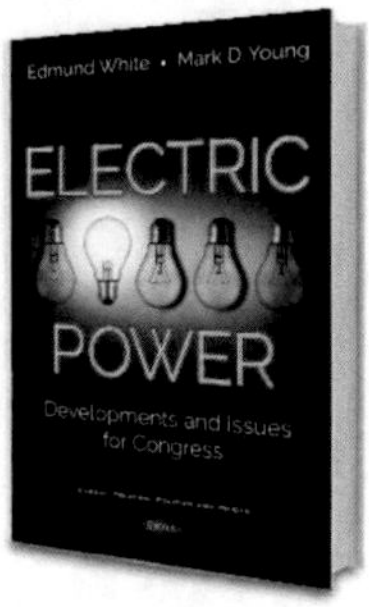

Authors: Edmund White, Mark D. Young

Series: Energy Policies, Politics and Prices

Book Description: This book is a compilation of CRS reports on electric power. The large-scale damage caused by Hurricane Irma and Hurricane Maria is examined in the context of policy options Congress may consider in order to help remediate such damage to the electrical grid in the future.

Softcover ISBN: 978-1-53614-357-7
Retail Price: $82

Energy Policies: Issues and Legislation

Editor: Doug Higgins

Series: Energy Policies, Politics and Prices

Book Description: This book is a compilation of CRS reports on energy policies. Some topics discussed herein include the energy-related research and development of coal-based synthetic petroleum and atomic bombs, oil- and gas-related wells are, the Atomic Energy Act, the Federal Energy Regulatory Commission, energy tax provisions, the Clean Power Plan, Section 201 safeguard actions and the Low Income Home Energy Assistance Program.

Softcover ISBN: 978-1-53614-485-7
Retail Price: $95

Energy Policy: Perspectives, Challenges and Future Directions

Author: K. J. Sreekanth, Ph.D. (Energy and Building Research Center, Kuwait Institute for scientific research, Kuwait)

Series: Energy Policies, Politics and Prices

Book Description: Sustainable development is triggering a re-assessment of innovation and technological change in all fields, and energy is no exception.

Softcover ISBN: 978-1-53613-744-6
Retail Price: $95

Energy Security: Perspectives, Improvement Strategies and Challenges

Editos: Mirjana Radovanović (EDUCONS University, Faculty of Security Studies, Sremska Kamenica, Serbia)

Series: Energy Policies, Politics and Prices

Book Description: This book presents the most important aspects of energy security, including its role as an element of national security and a factor of geopolitical stability. Many factors affect energy security (and vice versa) in different ways.

Hardcover ISBN: 978-1-53613-508-4
Retail Price: $160